Venusian Seed

Venusian Seed

Stephen Skinner

ATHENA PRESS
LONDON

Preface

This book was written with the help and guidance of my friends and loved ones in the spirit world. A special thanks to my guide and close friend, whom I call Mark. The purpose of this book is to try and break down, in some small way, the deep, dense, thick cloud of ignorance which has formed at this moment in time around our physical world. This ignorance penetrates all walks of physical life, to a greater or lesser degree, depending on the individual soul. If anything that I have written touches the soul of just one person, then this book will truly have been well worth writing.

Dedications

This book is dedicated to Mr D Skinner, my father, now passed on to the spirit world. He gave many years in service to the spiritualist movement, as a platform medium. Also to Mrs E Skinner, my mother, whose love, patience and understanding brought me to a deeper realisation of love. Between the two, they gave me a good, strong spiritual foundation, so that I may, God willing, continue the work that I am doing. And finally, to Miss S Skinner, my sister, whose typing abilities I could not have done without.

In the Beginning

Total darkness, no light. Nothing exists within darkness. Whether it be of the spiritual or the physical. Light is life, formed out of darkness, over many billions of years. All that was first formed in the spirit world before it entered into matter. First, sparks of pure spiritual living energy formed. It was the only light or life in a darkened universe. This speck of light would have formed out of the density of darkness. Evolving slowly but surely, at each step, towards a brighter level, then the level it had just formed out of, evolving over a long period of time, through the full spectrum of colour. This pure light or energy is the basic building material for all that is, on all levels, from the lowest to the highest. The stronger energies of light will always attract or influence the weaker, and will draw these towards itself by giving light or life to the weaker through its own power, knowledge and understanding. This is the law of all things within the spiritual and physical universe. These laws are positive in action and will continually repeat themselves in order to evolve or advance forward, to a higher state than they once were.

Laws do not change, they advance forward, only because of a deeper understanding of the original law. Change, as in progress, only comes about when a power or influence has reached its limit, and cannot produce any new forms of life out of itself. When it can only reproduce itself, this creates a state of stagnation. All things need to go forward to survive. To truly go forward, the power needs to draw other powers, different to itself, towards itself, to produce many new forms of life. In this law, the stronger power feeds the weaker, which enables the weaker to evolve to a stage where it can produce power for itself. Part of this power is sent back to the source, the higher power above itself. In strengthening the weaker power, and allowing that power to take a step further on that spiritual evolutionary ladder of life, enables the stronger power to give more influence to the weaker.

It strengthens the weaker to such an extent that it can feed power to that which is weaker than itself. The weaker will always need the power from the stronger, the stronger will always feed from the power of the weaker. This is law. As the influence of the greater becomes stronger, it draws to itself the weaker, eventually immersing it with the stronger, producing one power – the sun. Power attracts that which is like itself, or close to itself, because that is all it knows to attract. A continual learning process, expanding the power of itself, at every step it takes. To understand this first law gives a person a deeper concept of understanding, of how our universe was created, and a greater awareness of the power of the creator.

Physical Life

As power first formed in spirit, it would only be a short time before it formed into physical matter – into the world we know. The first spiritual forming of power into a physical world, would have been air, gas and then water. At the stage when a gaseous state formed, is produced every possible combination of itself, through trial and error. Some combinations worked together, others did not. Those that worked became law, and would be used and refined continually. That which did not work would reach a level, as far as it could advance, then it would disperse from that level, and would become a part of the level below itself again. If a system or power cannot advance, then it loses its purpose for existence on that level. Everything has a purpose; without purpose, it cannot exist. Everything that happens, whether it is positive or negative, is registered within itself, creating a pattern or plan of its own personal existence. This is called karma, and everything that is, has a karmic plan. Karma is a pattern of every individual life form, the plan behind all that is in existence. Although the basic pattern throughout all systems should be similar, any slight change would produce different combinations of life, and so produce different gaseous states. These gaseous states would flow freely through the universe, because there is no controlling force or influence to keep them in a set orbit.

Like the sun's influence on the planets, everything in the physical world has a spiritual counterpart, the spiritual being what goes on after physical life. The universe as we know it was produced as follows. Two different gaseous states reached a level where both could not advance any further. Both powers going down two different evolutionary pathways were attracted towards each other, by the necessity of advancing. Flowing into each other, one ignited the other, producing a mega-powerful explosion, lighting up a darkened physical universe and scattering out into space many thousands of new forms of life; new life formed out of the combination of the two powers. Some combinations of energy would be very quickly influenced by the stronger forces, back to the first law. Some would be blasted far into space, totally beyond the influence of any stronger power. As stated, any energy needs to feed off lower forms of energy, to survive. The stronger energy, with its greater influence, would form into a sun. Everything attracts the same as itself, to itself. At this time, no solid planets would have been formed, because to create the right combinations of gases to produce solid matter takes time, whereas an explosion is instant. Those furthest away from the explosion would draw to themselves that which was alike, and also powers that are lower. The combinations of life, caught in this power's influence, could not be used by that power. There would be forms around the outer surface, building up combinations of life forms for the power's future development, eventually becoming the Earth's crust. Sooner or later this power would be influenced by a greater power, the sun. With the influence of the sun, which creates activity within, are powered gaseous states, creating matter. The greater the influence of the sun, the greater the activity in the weaker power.

Our System

After the creation of solid matter, many different types of gases formed, underneath the Earth's surface. Eventually, they rose out onto the surface, because of a strong build-up in pressure caused by the gases continually producing themselves. Breaking through at the surface's thinnest point and producing volcanic eruptions within the earth, were created tunnels, because of the volcanic flow, burning away all the thin, loose rock, trying to force its way to the surface. After these tunnels had cooled down, water rose up, from deep within the ground, through the tunnels the volcanic eruptions had created in its flow. The water shaped the tunnels into great massive caves. In these caves, great lakes formed. All life was first formed out of these lakes. Charles Darwin was correct, as far as he went; he was a master of his time, who could only understand the physical world. If he had taken his theory one step further, to a spiritual evolutionary pathway, instead of just a physical pathway, in old age he might not have been so confused when trying to work out the physical pathway in the Bible teachings, for the Bible is largely about the spiritual pathway of humankind.

Life forms out of water, and before Noah there was no water on the surface. Thus, all life must have formed from the water in the caves. Plant life formed in the caves, carried out onto the surface by small animals and birds. The small amounts of water that had formed on the surface were quickly dried up by the sun. All bird life formed from bats, and roots needed to be strong enough to get deep down into the ground to search for water. Animal life forms, that had evolved in the caves, eventually came onto the surface, which was now covered with trees and plants. The human race formed from apes, but without any direction. Aimlessly, they roamed the Earth, with no other purpose except to survive and propagate its own species. On the planet Venus, the last planet in our system to be inhabited by human beings before the Earth, it was getting too hot to survive because all planets are drawn into the sun eventually. In Revelations, the last chapter of the Bible, it is written that this Earth will end in fire. They, the people, needed to escape from their burning planet. Only two ways were open to them, both dangerous. One was to travel through the vortex, to a parallel system, which I will write more about later on in this book. The other was to come to Earth. They split into two groups. In the Bible, it is written that before Noah, the gods came down from the sky, looked upon earthly women and took them for their mates. The earthlings were like wild animals, at this point only two or three steps away from inhabiting trees. They had only just learned how to stand upright, and use simple tools made out of animals' bones.

The Biblical Genesis

The Venusians found life on Earth very difficult. Entering onto a lower, denser planet than they where used to, and breathing the air of a denser atmosphere, must have been like continually taking in the air of a thick fog, creating many breathing problems. Disease would have also been a major problem seeing that the Venusians' immune system was very weak compared to the earthlings of that time. Whatever disease there is, the body's immune system will eventually conquer that disease. Any problems that physical life can offer, the human body is, within reason, built to cope with it. This shows us, and gives us a greater awareness on how magnificently we are built, and the power and majesty of our creator, our father God. Disease is strongest on the level of density that it is on. The more pure or refined the density, the lesser the disease. A time of great density produces a time of great disease, and a strong immune system to cope. If a race of human beings has become so advanced that on its own planet it had conquered disease, then the immune system within this race would have no purpose, and so would become weak with the lack of use.

However, now they were wide open to all infections. The Venusians, being of a higher intelligence than their earthly younger brothers, knew that their past was the earthlings' future. They tried to teach their younger brothers and sisters a way of peace. They brought us a dream, which for them was a reality, because it was their past, and the Earth's future. This dream is in symbolic form, and is the story of Adam and Eve. Paradise is a future state of humankind. The Earth has never been a paradise; paradise means a place opened up, within the light. The Venusians came to teach a darkened world about the light. Two great powers are symbolised by the story of Adam and Eve. The providing, physical creation of Mother Earth, and our spiritual creation of our heavenly father God. The masculine forces in nature, from above, and the feminine forces of nature, from below, symbolised as Adam and Eve in the Bible. This being the start of both powers working up eventually to meeting on the same level of thought, on the Earth, this happened many thousands of years later, in Jesus Christ's day, which will be written about in greater detail later in this book. They, a people without knowledge, which is symbolised as nakedness, were those without shame. The trees represent the Venusians, the bearers of knowledge. The apple represents Venusian knowledge and understanding. Eve took the apple off the tree first, symbolising a strong female influence on the Earth at that time, because the female can give birth, and the great Mother Earth provided for all her children's needs. With Venusian knowledge, female influence grew into a great power. Some followed the true teachings of the Venusian gods and others used that power to control and possess everything they desired. Putting such great knowledge in the hands of an undeveloped race was like giving a loaded gun to a child. They built great statues of their female goddesses, giving them attributes of the first Venusian generation on Earth, and worshipping them. Knowledge in undeveloped minds brought chaos, confusion and much bloodshed.

The earthlings' inter-breeding with the Venusians produced within the children a higher intelligence than that which its earthly parent possessed. However, it would still have a lower intelligence than its Venusian parent, which would eventually totally wipe out the Venusian seed. The children of these mixed relationships, with their heightened level of intelligence, with partly understood but undeveloped Venusian knowledge, had not yet the strength of mind to be able to cope with the power that this understanding would produce, thus sending the earthling into a darker stage, than they were at before the understanding of Venusian knowledge came their way. The Venusian seed had to survive, for the future development of these children, into the light, or they would have been lost in their own depravity. So the Venusians cut off the earthly human race from any further Venusian knowledge. Together as one pure Venusian race, they moved away from their younger brothers and sisters, into the mountains, where the air was easier and better for them to breathe.

The serpent in the Bible represents the desire to know more, which in itself is not a sin. How they used that knowledge is what brought about their destruction. Slowly, out of the female influence, a stronger more powerful male influence formed. This is symbolised in the Bible, as Eve giving the apple to Adam. Being thrown out of paradise represents being cut off from Venusian knowledge, or from the gods of that time. Eve had two sons, symbolising the positive and negative powers of the Earth. One brother killing the other showed what state the Earth was in: the negative power was stronger than the positive, as shown in the story of Cain and Abel. At this point in time, it had not rained on the Earth. The Venusian gods did not leave the earthly human race to its own fate. Those that practiced the Venusian teachings, and used the knowledge that they had been given correctly, were guided away from disaster, because they had the Venusian seed still burning brightly inside of them. That seed needed to be protected at all costs, and one family that followed the true way was Noah's family. Whether Noah went into the mountains to be instructed on the building of a boat, and also told about the coming disaster, or if the gods came down from the mountains to Noah, we do not know. Noah *must* have had Venusian knowledge, as water had never been seen on the surface of the earth, so how would he know to build a boat, and the knowledge to go about it?

The Flood

The Earth at this time was very unstable. After the build up of gases had broken out onto its surface, water formed underneath, and because of the build-up of water pressure, water broke out through the surface's weakest point, creating a major earthquake. In this earthquake, part of our Earth broke away. The space left was quickly filled with water from underneath the Earth. This part that broke away, became our moon. Scientists have found that rocks collected from the moon were the same as rocks in the area from where they think the moon broke away. Thus, there is an ocean on Earth that is so deep that the moon could fit into it.

For water to flow on the surface of the Earth, it needs a controlling force, and that is why our moon was created from the Earth, at the time it was. When anything has a purpose, it will come into being. The sun dried up part of the surface's water, creating the cycle of rain. The Venusian gods survived the flood, they may have built their own ships, or maybe it was because they lived in the high mountains. There is only one strain of human life, formed from the ape. Many different forms of animals, attempted that next step in evolution, but they did not survive the flood, thus strengthening the one human seed, the only seed left. The ark was a fairly small boat, compared to the ships we can build today. Very few animals would have been necessary to keep the animal kingdom alive. A pair of cats, for example, covers many forms of animals developed out of one pair. That is how all the species could get into such a small ark. The missing link between apes and human beings was covered over by the sea, at the time of the flood. The day is coming when we are drawing towards the end of the Piscean cycle, and entering into the cycle of Aquarius, which starts around about the year 2000. The Earth will become more active. You will be able to see predictions of the past, working themselves out in the present the deeper we enter into the Aquarian age, predicting a future from the actions of our past. Some areas that are now sea will be land. This will bring to the scientist much knowledge of how the human race formed. Everything that is evolves – even the earth itself.

The Tower of Babel

After the rain had stopped, the earthlings multiplied themselves, becoming a great power. Not satisfied in worshipping God, they thought themselves to be gods. They tried to bring down anything that was beyond their level of understanding. They tried to destroy the source of where all earthly knowledge comes from, that is, the Venusian gods. This is symbolised in the Bible, in the story of the Tower of Babel: one nation, one negative destructive force, all speaking the same language, worshipping themselves as gods. The Venusian gods had to control this great mass of destructive power. They needed to teach earthly humankind a lesson. They could have destroyed the people, who were turning against them, but they did not. What they did was send a space-craft to hover over the tower, and sent a powerful beam of energy towards it, smashing the tower down to the ground, producing fear and confusion in all who witnessed it. It was a lesson to be learned, but only the tower came to destruction. All the gods are loving and caring, and do not bring unnecessary suffering on the earthly race. A good percentage of suffering the human race bring upon themselves.

This incident scattered the people, who were never again to be united as one, as could create such a destructive force. The distances between the many small groups, slowly changed the language. Although all language came from one language, possibly Latin. If you have two towns close to each other, there will be a slight difference in their language. Think for a moment of the change in language over thousands of miles. As the human race wandered the earth, language changed to such a degree that it became very hard to hear the similarity between one language and the original language.

Abraham

Two great powers now ruled the earth; male and female religious beliefs. The female worshipped nature, the male worshipped the sun, moon and stars. One great nation in the beginning, but because of the greed for power, and two different belief systems, they had to part and go their separate ways. If not, they would have been at war with each other. Rome became the centre of the male influence, Egypt the centre of the female influence. Both powers went down negative pathways. The Venusian gods had to search over all the earth to find a race of people that had gone down the true positive pathway, as taught by them. Most people of the Earth had so corrupted themselves that they were not fit to carry within themselves, or give birth to, the pure Venusian seed. Only one man was found to have travelled down the true pathway. His name was Abraham, the father of the pure Venusian seed. The seed needed to be protected, at all costs. Many of the Jewish laws in the Old Testament were created for that purpose. Circumcision was a sign, a way of separating all other nations from the pure race along with the law that they could not marry out of their faith. These laws are not necessarily needed in our modern times – in fact, they have not been necessary since the birth of Jesus Christ.

To bring forward any change in the evolutionary pattern of life, many thousands of souls in the spirit world, and also in the body of flesh, are needed to bring this progress into being, and Abraham's role was so important, in the advancement of human-kind. He needed to be tested to the full by offering, as a sacrifice, the life force of his only beloved son, that he had had with Sarah, his wife. He was proved to be true to the light, and pure enough in spirit to be able to take on the role of being the father to many nations. Before his son was born, three Venusian gods came down from the mountains where they lived and visited Abraham. They drank wine and ate bread, as only spirits do not need physical sustenance – physical bodies do. There is no spiritual meaning to this action, except to say the gods of that time were in physical bodies. The point of this story is the message, given to Abraham from the Venusian gods. This was that his wife, Sarah, would give birth to a son, although she was in old age. Any true God is of the light. No darkness or negative thought or action can be a part of the true pure life force, that flows to all from our father God. God can only create, it is not within his make-up to destroy. If he could destroy then how could he be a God of love and compassion? So looking at the Bible with open eyes, it shows us that the destruction of Sodom and Gomorrah was a natural phenomenon. The Bible says, 'The heavens opened up, pouring out fire and brimstone on the two cities.' A burning meteor would have the same effect. The Venusian gods knew what was going to happen. They tried to warn the people of the coming destruction. Lot's family, who lived in Sodom, heard the warnings sent to them by the gods. Lot needed to be tested, to see if he had been affected by the gross darkness. Otherwise he could not go back into the race of people of the pure seed, because he had corrupted himself, living amongst such people.

Instead of offering up his son as a sacrifice to God, as in the testing of Abraham, Lot offered up his two daughters. No harm came to any of that family, because of the faith of Lot, to his God. Because of the negative way the people of Sodom and Gomorrah lived, their lives had made them deaf and blind to the truth, this leading to destruction. No matter how low a person becomes, God will always try and help them, but only if they allow him to. A question to the reader: do we listen to his warnings or are we deaf and blind, just like the inhabitants of Sodom and Gomorrah?

God never turns away from his children, we turn away from him. For centuries the Jews and the Arabs have been at war over land that both say belongs to them. Abraham was the uncle of Lot, Lot is the father of the Arab nation. Abraham is the father of the Jews, and both nations stem from the same seed. Terah, who was the father of Abraham, showed that the war between the Arabs and the Jews was brother fighting brother, because they both came from the same family. In King James's Bible, Genesis 13, Abraham and Lot part company. Abraham and Lot were very wealthy in cattle, the land could not feed so great a herd. Abraham said to Lot, 'This is a great land, before us, if you wish to take the left I will take the right.' Lot took the Jordan Valley as his home.

Jacob and Family

Every person that lives a physical life, lives that life to progress and go forward. A plan or a pathway of life is created for every individual, before he or she enters into the womb. We may wander from our own personal pattern of life, as we all do, at some point in our lives, usually because of suffering, brought about by lessons that need to be learned. But we will eventually, through the necessity to survive, go back onto that pre-conceived plan. Which was created for our personal advancement, before we were born. Jacob turned away from his family, and his God, for monetary gain by swindling his own brother out of his birthright. Although he had turned away from his creator, God did not leave him to wander aimlessly in the dark. Instead, he showed Jacob a pattern of his universe in a dream.

Jacob was sleeping in the open one night, with a stone for his pillow. His mind was opened to great spiritual truths while he was dreaming. (People are closest, to the spirit world, when they are asleep.) To understand dreams is to understand many deep spiritual messages that are sent to guide us, from our spirit loved ones. In sleep, we can also discover many deep feelings and emotions within ourselves that we keep bottled up, but which we should bring to the surface and release. Negative emotions block the flow of life power that comes from our father God. Jacob dreamed of a ladder, stretching from Earth to heaven. On that ladder, angels, or spirit people (the meaning is the same), ascended and descended. Both were messengers from the light, showing us that every rung of the ladder symbolises plains of thought. On every plain, spirit people live, brought together by similar thinking and life patterns. The plain of thought, that they live in, is equivalent to their spiritual advancement, within the light. There are many worlds, or plains of thought, and each rung of the ladder represents one level of thought.

The meaning of this dream is very close to what Jesus said in the New Testament, 'In my father's house, there are many mansions.' Because Jacob cheated his brother out of his birthright, he was also deceived by Rachel's father when he tried to make Rachel his wife. Whatever we do in life, whether it be positive or negative, comes back to us, as shown in this story. To a certain degree we create the negative in our future, by the actions we do today.

Jacob returned to his own family, to the brother that he had dishonoured, and he was afraid of his brother. Steeped in his own guilt for what he had done, he wrestled all night long with his own conscience, his troubled mind. Should he face his brother, which would be facing the responsibility of his own actions, or run away? Face up to any situation that comes your way, do not keep on putting it off until another day. His family was pleased to see him, as God is pleased to see any of his children who stray and return to the true pathway.

Joseph

Jacob had twelve sons, and Joseph was one of those sons. He was also the son of Rachel, who was greatly loved by Jacob. Joseph walked the way of the Lord, and the power of God was with him. A great natural disaster was about to threaten the pure Venusian seed within the souls of the Israelite nation. A great famine was to cover the land, and Joseph was the only man who was able to save, and preserve, the spiritual seed of Israel. The Venusian seed, and the seed of Israel, which is one and the same, had to survive at all costs. As with Jesus, his own people, who were of the village that he was brought up in, did not believe in him and turned against him. So Joseph's brothers turned against him. His brothers abandoned him in a pit. Then they put him into slavery in a foreign land, for money. Imprisoned on the word of a harlot, he was tested to the limit. His belief in our father God was still strong – what faith he must have had! For anyone to be chosen by God, it is because of the level of spirituality that they have reached. Then, they need to be tested, to see if they can achieve the task that God had planned for them. For anything to happen in a physical world, many thousands of souls are involved – in both the physical world and the spirit world. The greater the suffering we have in life, the greater the glory God as planned for us. We can only be tested through suffering. Joseph eventually took his people to Egypt, away from certain death and famine. The pure seed needed the knowledge of the mother, Egypt, but did not need to be a part of her influence, because she had corrupted herself, and was walking down a pathway to certain destruction.

Exodus

Great sorrow had fallen on the people of Israel, at the hands of the Egyptians. A leader needed to be found, amongst the Israelites. God chose Moses to be that leader, to bring down the great and powerful nation that Egypt was. Moses had to be brought up by the Egyptians so he could learn their ways, their strengths and weaknesses and most of all their knowledge which the Venusian gods had left with the Egyptians, many hundreds of years before this point in time. Most of the Pharaohs had Venusian blood running through their veins, so did the Roman Emperors. In Genesis, chapter six, verses one to three, it is written that the sons of God saw that the daughters of men were fair, so they took them for their mates. Where could these sons have come from except from the burning planet Venus?

The Israelites had very little knowledge of their own, before entering into Egypt. A belief in a creator, this knowledge given to them by the Venusian gods, a knowledge of a belief system, is not the same as a knowledge of technology, like the Romans and Egyptians had. The Israelites knew how to farm the land, and look after cattle. They learned all the knowledge that they needed from their Egyptian masters. There was a group, called the inner circle which was built up of Egyptian priests. This inner power of knowledge was kept in the priesthood. Only a Pharaoh or one of his family could learn this knowledge. The Pharaohs did not acquire this knowledge – they trusted totally on the words of their priests, who like the Egyptian nation, had also defiled themselves. This is why Moses was brought up in an Egyptian household to find the source of their power, then use it against them, which he did. The burning bush, where Moses communicated with the Venusian gods, was seen as a miracle. What he was really looking at was an illusion created by a laser beam, from a Venusian spacecraft, a similar idea used today at pop concerts. The voice of God would have been a loud speaker system, also coming from the spacecraft. Moses was kept at a distance when the gods were leaving, and became scared and hid his face, so he did not see what was really there. He saw a strong light, which could have been the spacecraft taking off. The mind is a powerful instrument; believe strongly enough that something will happen and it will. Moses believed all that was told to him, and what he thought he saw. In Moses's case, he believed his healthy hand would look leprous because he believed what his Venusian God had told him. Some Catholics believe the crucifixion of Jesus so strongly that the wounds that Jesus had appear on their bodies. Same as Moses's hand – powerfully strong belief. To turn water into blood, as Moses did, could be the same as changing water to wine, as Jesus did. Venusian intelligence could have taught Moses how to heighten or lower the vibration level on anything, creating a change in the molecular system of that thing, thus changing it to something different than what it once was. This is how Jesus did it. The rod to a serpent, the same, a change in the molecular system. When it comes to the Nile becoming the colour of blood, this could have been a natural happening. Earthquakes were common in the east and are still so today. A small earthquake dislodging a massive amount of red sandstone, which is very common in that country, broke into the Nile, which would make the Nile the colour of blood, and also would give it a bitter taste.

The Nile was so polluted by the red sandstone that the frogs came out of the Nile and onto the shore. The frogs, by drinking the water of the Nile, poisoned themselves and died a few days later, seeing the Nile had been so fouled and the Egyptians were not a clean people, bodily or morally unlike the Israelites. The Israelites lived away from the Egyptians and probably had a hidden fresh water supply. Gnats and flies are drawn to filth, so it was the way the Egyptians lived that brought these plagues upon them. Because of how the Israelites lived, this is what protected them from the plagues. A strong wind could have easily moved the gnats and flies out of Egypt. The cattle may have drank from the fouled Nile, or become diseased from eating off the land, because the land had also been fouled by the frogs, gnats and flies. From the fouling of the Nile a pattern seems to be emerging; the reason for all the plagues started with the fouling of the Nile. If the Egyptians drank of the Nile and ate poisoned meat, this could have created many different illnesses, boils and blisters, being just two of them. A high powered electro-magnetic force hitting a specified area in the atmosphere above Egypt would have caused a dramatic change in the weather. As with rain and hail stones, the locusts are brought in on the wind and take away on the wind. Nothing can ever happen unless the conditions are correct for it to happen. This goes for positive or in the Egyptians' case, the negative. A chain of incidents because of the fouling of the Nile, including all of the plagues, covering fouling and diseased water and food and the effect on human beings, could easily have been a natural occurrence. The darkness that covered the land of Egypt could have been an eclipse of the sun, by the moon. The Israelites had previous knowledge from Moses, which Moses got from the Venusian gods. They got themselves prepared, that is why there was light in the Israelites homes and the Egyptians were in the dark. The people of Egypt must have been thirsty, because of the fouling of the Nile, hungry because all food, whether it be meat or cereal, was destroyed or badly fouled with disease. Feeling low, confused and scared, even doubting the power of their own gods, they must have wondered what was to come next. Many died that night, the old, the very young and the weakest. This was not a personal thing, as the Bible tries to portray. Because the God is a God of love, and destruction is not in His nature. Disease takes all the humans it can; the strong survive, the weak die. There probably was not a household that had not known death, but not all of them were first born. In the Bible, the span from the beginning to the end of the plagues seems a very short time. I would have thought that it would have happened over many years. A slow but sure deterioration that the Egyptians had brought onto themselves.

The Red Sea

The Gods were with the Israelites, a pillar of cloud by day, a pillar of fire by night, created from a Venusian craft, which hovered in the clouds above them. The Venusians were ready for battle against the Egyptians. This would have been a last resort, if all else failed. At the Red Sea the Israelites walked on dry land, and the sea parted for them. Moses was told by the Venusian gods what was about to happen beforehand. Further up land there was an earthquake which opened up the ground. The water from the Red Sea flowed into the hole that had just been created producing dry land for Moses and his people to walk across. Another earthquake closed the hole, leaving water pouring out onto what was once dry land, drowning every man in the Egyptian army who tried to follow the Israelites.

The Venusian gods supplied food and water to the Israelites while they were in the desert: the water from a stone, quails and bread flakes, called manna. They distributed this food from their spacecrafts. Today, as we enter the twenty-first century, UFOs seen frequently in our skies have many times throughout our history left behind them bread-like flakes, which seem to dissolve with a touch. The manna in the Bible seems to be of a stronger version, because it lasted longer. The manna from the UFOs of today is called Angel Hair.

At Mount Sinai the Venusian gods spoke to Moses. The mountain was wrapped in smoke, the Venusian gods descended onto the mountain in fire. As you would get with the landing of a spacecraft, thunder and lightning, and a possible strong build up of electro-magnetic power created from the spacecraft, or the movement of that craft. The gods gave Moses the Ten Commandments, the law of the temple and the instructions on how to build the temple. This was done to centralise the Israelite nations around the temple so that they would not wander and corrupt themselves with other nations' negative beliefs, so as to keep pure the Venusian seed that had been bred into the Israelite nations.

Aaron's family was chosen by the Gods to be the priests and caretakers of the temple. Aaron's trade was a worker in gold, the only person that could have created a golden calf while Moses was talking to the gods. The people went to Aaron, who was a craftsman in gold and was also Moses's brother, and said, 'Create us a God,' and he did – a golden calf. Moses was in the mountains a long while. It seems Moses's brother Aaron wished the power and leadership that Moses had for himself. By creating the golden calf for the people of Israel to worship, Moses still gave the priesthood to Aaron's family, although Aaron had turned against Moses and his God, by attempting to corrupt the people with false gods, such as the golden calf. One of the Ten Commandments says, 'Thou shall not kill.' Moses, as a young man, killed an Egyptian, leaving on his hands and in his heart the bloodguilt for his actions. We all make mistakes, and we should learn from them. Moses multiplied his mistakes by ordering the deaths of three thousand of his own people in Exodus, chapter 32, verse 25 to 29, 'Who is on the Lord's side…' Moses suffered greatly for his actions, and he was not allowed to enter into the Promised Land.

When Moses came down from the mountain and saw what had happened he was very angry with the people of Israel. For some unknown reason he was not angry with his brother Aaron who brought about the corruption of the people by his betrayal of Moses and the Venusian gods. It is no wonder what Jesus said about Aaron's priesthood: 'Do as they say, but not as they do, the truth is not within them.' This was the start of Israel coming together as one nation. Protected from corruption by the strictness of their laws and separated from other nations so the pure seed could survive. Most of the Old Testament is the history of the Venusian gods' influence on the Israelites, allowing them to wander for a while then bringing them back onto the right pathway of life. This is also the history of the pure seed, from its conception to its birth within the physical world. The birth of the pure seed is the birth of Jesus. The pure Venusian seed, manifested into an earthly body.

THE VOID

The Void

As a solid person, living in a physical world, all that I see is solid. But between two solid objects like the Earth and Venus there are many levels of spiritual consciousness. As these levels descend, from the creative force which we call God, they fall down to the earth plane as far as they can go before becoming enveloped in negative vibrations which is the darkness of this earth. The earth's spiritual consciousness rises upwards towards the consciousness from above. There is a gap between the two, the highest from above and the lowest from the earth. The gap in-between is called the void, whether it be of a spiritual or physical nature. The Venusian seed, born into earthlings, was the only bridge or channel to our Father God. Before the time of Jesus Christ there was no link from the earth to our Creator. The Venusians entered onto our world to sow the seed. That would eventually close, for all time, the void. In so doing, it allowed the light to shine in our dark, negative, earthly world. That which flows from above needs to be one with that which flows upwards from below. From the time the Venusian gods came onto our world to the time of Jesus, the seed needed to be protected. That is what the Old Testament is all about, protection of the seed. If the seed had become lost in the negative earthly forces we, as a people, would still be in the Dark Ages and Jesus could not have been born. The pure seed that was Israel, needed to have knowledge of the mother Egypt which was brought about by the great famine in Joseph's day. Rome, the father, ruled the Jews at the time of Jesus, so Jesus was brought up under the influence of Rome. The pure seed would have had knowledge of mother and father. The pure seed was many times scattered. The gods, by sending messengers to the Israelites, who were the prophets, directed the people back to being the one pure race. By the influence of the masculine and the feminine in the womb of Israel, a child was born who was able to bridge that gap between humans and their Creator.

Biblical Proof

We are spirit, who have entered by our own free will into a physical body, for the purpose of learning. Genesis, chapter 1, verse 26: 'God made man in his own image.' If God was physical we would see him walking amongst us. God must be spirit and if he created us in his own image we must be spirit also. God is love, we are in a physical body to learn how to practise all concepts of that love, that flows from our Heavenly Father to every individual soul. If the Venusians did not exist then who told Abraham that his wife was with child, in old age? Who taught Noah how to build a boat, when there was no surface water and thus no need to know how to build a boat? Who had the knowledge to build the pyramids or the temples in Egypt or Rome? Earthlings on their own without outside knowledge who were just experimenting with very basic tools, with their evolutionary level about four or five steps beyond the ape, suddenly able to build pyramids? The human, Earthly race did not have time to develop to such an extent, to be able to create such wonderful powerful buildings. The void separated us from our Creator so that only Venusians that could have taught us. Because of how the Earthlings used the knowledge, negatively, for power and control of others so as to gratify their bodily needs and base desires, the Venusians parted from the earthlings and entered into the high mountains. It was easier for them to breathe the more purified air in the mountains.

The Ten Commandments were given to Moses on Mount Sinai. In Greek mythology the gods lived in Mount Olympus. Mountains are mentioned all the way through the Old Testament as a place for earthlings to communicate with their gods. Caves that were opened up around about the early 1960s were found to have drawings of men in spacesuits painted on the walls, with a few paintings of the animals that were around at that time. These caves had been closed off from the human race for thousands of years. The time period was discovered with the animal paintings and the primitive tools that were dug up in the caves. We think of space travel as quite a modern phenomenon. If this is so, how could these cavemen have depicted so accurately the space suits, unless they had actually seen them? In the Sahara you can see from an aeroplane runways burnt into the ground with much use and from a powerful heat source. A powerful heat source that we still do not have today. This could have been a landing place for spacecrafts. The buildings of Rome and Egypt, and anywhere else on the earth the Venusians had settled, had to be big so as to let their own kind know where they were, and for a future time. The buildings had to be sturdy and built to last so they could be seen from space and because of the great amount of pure Venusian knowledge painted on the Egyptian pyramid walls, knowledge for future earthly generations. The mathematics in the structure of the pyramids, on their own, baffle the human race even to this day.

The wall pictures were drawn inside the pyramids to show earthlings how to use the power source the pyramids contained. The function as they were planned is not fully understood by humans and cannot yet be copied, but as we become more spiritual, the pyramids will give us greater information. Knowledge is not given to the human race until that race is ready and can understand that knowledge. All that is hidden, will be brought out into the open. All will be known. If knowledge is given before the human race is ready for that knowledge, they will abuse it as they have done before. True knowledge is hidden until the earth is spiritually ready. We have now passed the year 2000, the age of Aquarius, the age of enlightenment. We think we are wise, with much knowledge in technology and in medicine. But if all knowledge was symbolised as a bottle of milk, then all the knowledge that the human race possesses is but the cream. We have still got a long way to go.

Egyptians

Rome and Egypt were mixed with Venusian seed and earthly human seed. They believed what their older and wiser Venusian brothers and sisters had told them of the treaty – that they would return and take those of the Venusian seed back home with them. The pyramids were so cold inside they could preserve the still living yet sick bodies of the Egyptians in a deep sleep until their brothers returned. A lot of care was taken in the preservation of the body and oils: gold jewels and even slaves were placed in the pyramids for their use when they awoke.

Not so many years ago a pyramid came to the surface of the desert after being covered in sand for hundreds of years. This is why it was not found sooner. They took the mummy out of the pyramid and dissected it and found that a few cells in the mummy were still alive. This shows us that the pyramids were built for the *living* not the dead. A dead body quickly decays and turns to dust. If the pyramids were built for the living, which I think they were, they needed to be able to stand the test of time. They also needed to be secure enough not to be opened by anyone except by those who had knowledge to awaken them: their Venusian brothers and sisters. Thieves after gold and jewels, plus historians and scientists in ages past broke into the pyramids, changing the atmosphere within and bringing the temperature inside up to the same as the outside and bringing the Egyptian mummies, with Venusian seed bred into them, back to a full awakened state. The stories of mummies walking are quite common in folklore. Although grossly exaggerated there is still a powerfully strong grain of truth in the original stories. The disease that they had been able to contain because of the deep sleep state would also become active at the awakening of the mummy, quickly bringing to an end all those that got woken up too soon.

The curses that were written on the pyramids such as, 'Who breaks the seal of this tomb, shall die' came true for a group of scientists. It is not a curse but a *warning of the disease that was within the mummy*. The disease was within the pyramid, contained, so it did not affect others, and to slow down the decaying process within a physical body which is what the Egyptians were put into a sleep state for. If a highly intelligent being mated with a much lower level of being than himself, like the Venusians did with the earthlings, this creates a disorder in the genes, producing a wasting disease. The pharaohs could possibly have caught AIDS, and it would be interesting to know what the scientists died of. The last remaining pharaohs did not even reach their teens. Places of great Venusian knowledge have been carved into stone showing up all over the world. Even in Britain we have our Stonehenge.

THE VORTEX

The Vortex

The true seed had already been firmly placed in the womb of Israel to bring into being the future Messiah. This was their only purpose for entering into a lower density. Coming onto Earth was over. The last of the pure Venusian seed who gathered themselves together as one nation now called themselves Atlantians, working desperately to find a way home. The vortex is a gateway into the next higher system to our own which overlaps our Earth, at the Bermuda Triangle. There was no vortex until the Atlantians created one. They used the weakest point between the two worlds, at the point where they overlapped. The power that would have been necessary to create such a gateway would have been greater than the bombs dropped on Nagasaki and Hiroshima put together. It has been said that they used crystals to harness solar power, so they could direct that power to the place where they wished to create the vortex. Many were destroyed in the attempt, but a lot more made it home to their own people, and to their own true level of existence.

To open the vortex today needs a strong build up of electro-magnetic power, which is produced at its wildest state in thunder storms. Even after they had created a way home they never forgot the earthly race. They keep coming back to check on the progress of the human race. Venusians or Atlantians, whichever name you wish to call them, are the same people still watching over their younger brothers and sisters. We see them in their spacecrafts, in the skies at night; we now call them UFOs. Since the moment that we learned how to use nuclear power for destruction they have been visiting Earth more frequently. If we in a moment of madness destroyed our planet, since we have the capabilities to do so, we would affect their world also, seeing as there is an opening between the two. Entering into the vortex, as many a ship and aeroplane has done, would be like going through the eye of a hurricane being sucked into another dimension in time.

Our earth is built up of power lines, which we call ley lines. These ley lines come from the North Pole to the South Pole. The other set of ley lines goes from east to west. At the point where two ley lines meet there is a build up of great power. We, as human beings, have built our churches and places of worship on these ley lines. UFOs possibly use the ley lines to recharge their spacecrafts before going on their homeward journey. The ley lines could be used as a form of direction into our world, or as a method of propulsion for their crafts, like a glider uses the wind. The power at the point where two ley lines meet could possibly be used to communicate between two spaceships.

Beyond the Vortex

There is a world of a greater, higher intelligence then ours: a race that has entered into the golden age many thousands of years ago. We as earthlings are just about entering into the golden age; they are our future, we are their past. Whatever mistakes they made, they are trying to prevent us from making by keeping an eye on us. As our sun influences Mercury so does every planet influence every other planet. The same system functions beyond the vortex with its own sun and planets, overlapping our world at the Bermuda Triangle. Everything is influenced by what is higher or lower than itself. Their world strongly influences ours. The universe, what we know of it, is like a spider's web, made of power lines, the vortex being a power line to another world. There are many suns. Our sun is influenced by a bigger and greater sun, expanding out on all possible pathways, lighting a darkened universe, a continual advancement of the light. As humans we see only a very small amount of what goes on around us. We are still rather a primitive race compared to our older and wiser brothers and sisters, beyond the vortex. Because of this power, within the vortex is a spinning movement. The outer casing of a spacecraft must be able to spin in the same direction as the power of the vortex, otherwise, going against that current of power, would break up the craft. The inner casing would have to be stable, where the space persons would live. I have included a picture to possibly simplify my statements on the vortex.

The straight lines are the power that any solid object has over another. The blank circles are planets not yet named. A small part of a vast network.

The Birth of the Messiah

The Venusian seed was in the Mother; she was a descendant of King David, not Joseph. Joseph was a craftsman and a good father, he only needed to train and supply for the seed, which was Jesus's bodily needs, although Mary had been tested to the full, by living through many incarnations, to bring her to the high spiritual level that was necessary for her to be able to give birth to Jesus. Mary was a virgin; she gave birth to Jesus from the seed of a Venusian God, called Gabriel. The seed was artificially injected into Mary: she had not been with a man sexually. Many beliefs that we still have today, they give greater importance to Mary the Mother, than to Jesus, a son of a God. The true meaning of a word can change drastically when going from one language to another. The reason for this is in the story of Paul when he was in Rome. The Romans worshipped gods and goddesses. They took great interest in what Paul was saying about the new Christian belief. They could easily see a God-like male personality in the form of Jesus, in Christianity. There was no female goddess to worship so they created a goddess in Mary. Many gave up their old pagan ways, and their pagan Gods, entering fully into Christianity. Some became Christians but still practised their old beliefs, hiding their own pagan rituals behind Christian holy days, changing from one to the other when the opinions of Rome changed.

The New Testament was first written in Greek, and the word 'virgin' in Greek means woman who has not given birth to a child. Different to what we understand the word virgin to mean. Mary was a very intelligent spiritual lady who taught her son the basic understanding of spiritual belief, and how to bring out within himself the power and understanding, so he could truly know the Father of Creation. In so doing, she gave him the power and capability which was necessary for him to do – his spiritual Father's work. A child needs to go through the natural process of learning. He would not have been born with the full knowledge of who he was. He would have known at a young age that he was different from all the people and children that he came into contact with. How different he was and what his true purpose in life was I do not think he fully knew or understood until the day he was baptised by John, his cousin. Where did Mary get the knowledge to be able to teach Jesus?

The only true places of Venusian knowledge were hidden from the people of those two countries, and were in Egypt and Rome, and both countries had blinded themselves from the truth by the negative way they led their lives. Jesus needed to be influenced by the Father Rome, which he received by being brought up in a Roman-conquered country. He also needed the influence of the Mother which was Egypt. Joseph, Mary and Jesus escaped to Egypt, to break away from the evil grasp of King Herod. This happened after the three wise men visited Jesus in a house (not a stable) two years after he was born. Joseph was instructed by a spiritual being in a dream to go to Egypt because the child was in danger from King Herod. Jesus and family settled just outside Egypt with a group of priests and priestesses, who called themselves the inner circle. They had been members when they saw how corrupt the priesthood and Egypt itself had become. Those that followed the true way broke away from Egypt and its priesthood. They settled down in a small group just outside Egypt. This is where Mary learned all the knowledge that was necessary to teach her son, Jesus. The star that guided the wise men to Jesus could have been a Venusian spacecraft.

Any solid object in space would travel towards the greatest influence, and to hover over a specific house would seem like there is direction, and behind direction – a thinking mind. Not a solid object but a Venusian craft. The shepherds were a very privileged people: they – of all physical earthly human beings over many thousands of years – saw the curtain, which shrouds our world, created by negative thoughts, words and deeds. The spirit world raised the consciousness of the shepherds above the negative dark shroud that surrounds this physical world, opening their minds to the beauty and the existence of Spirit. One minute, they were looking at a dark sky; the next they would be seeing a world of light and colour. They came to see Jesus on the night he was born.

Early Ministry

Jesus seemed to come to full realisation about himself, and also of the pathway that lay ahead of him, the moment he was baptised. The great flow of spiritual power from above opening in Jesus a full awakening of all his physical and spiritual senses. A flow of pure knowledge, all in a moment. This would have brought about great confusion in Jesus, like overpowering a computer with data. It is no wonder that he headed for the solitude of the desert. We are physical, like a machine controlled by a spirit, our true self. Good and evil is inside us. The creative power exists; if not, we would not be here. The Devil does not exist.

Jesus needed time to work out all the power and knowledge he had just received. As there is light and darkness created within the souls of all individuals on the earth, so there is in the spirit world. A collection of souls, like minds, whether good or bad, forms a plain on that level, the lowest being levels closest to the earth. Physical rises towards spirit, baptism or a wedding, as described in the Bible, is when physical becomes one with spirit, and physical matter loses its purpose, bringing an end to physical life. A wedding represents two becoming one. For the purpose of learning, there is physical and spirit. Once a physical person has learned all they can know in experiences, from being in the physical, then there is no purpose in taking on physical life; physical life has nothing to teach them. When two – physical and spirit – become one – a person then stays in Spirit. In Buddhism, this is called 'breaking the life and death cycle'. Spirit level is always one step above physical level. Jesus's spirit would have been very close to his physical, and the life he led would have strengthened the power in the physical towards his spirit. If he had not died on the cross, he would have died physically anyway because of the closeness of physical to spirit. From the moment he stepped out of the desert his ministry had begun. From the first Venusian thought of how to build a bridge, joining together one spiritual consciousness to another over the void, this was carefully planned, and put into motion in physical life, from the first step onto our planet Earth to the point of the birth of Jesus. Like a play with all the actors in place. A perfect plan with no room for error, because of its importance in raising the physical level of the Earth to a higher level.

With the birth of Jesus a new dawn had been brought into being. The light that shone through Jesus gave hope of a better tomorrow for all. Many of the old laws were no more. A change and renewal of the Old Testament's thinking came with the Venusian plan being perfected, and Judas Iscariot could not have been as the Bible portrays him. He was an actor playing his part, and nowhere in the Bible does it say, 'Jesus met Judas Iscariot.' It is written where Jesus met all his disciples except Judas. Possibly Judas was with him from the beginning, brought up in the same village. At the last supper Jesus said, 'He who will betray me tonight will put his hand in the meat dish when I do.' Judas was born to betray Jesus, otherwise that which has been written in the Old Testament would not come to pass. Judas loved Jesus and was about to fail him by not doing the task he had been given to do by God. But Jesus said, 'Go do what you have to do.' Suicide is a great sin, but not for Judas who hanged himself. Because he had done his job it was not necessary for his physical life to carry on. He could not learn any more from that level of earthly life. Jesus, who so loved his people, gave himself totally by willingly offering his life as a sacrifice for all, on the cross. We are all on different levels of development, and we all one day will reach the level Christ has been on. Jesus says this, when he says, 'Why do you marvel at what I do and say, there will be a time when all will be able to do as I do. Even greater things than what I do.' Jesus spoke many times of the golden age, the age that we will soon be entering into. Jesus commanded us to follow him and try to liken ourselves to him in all we think, do or say.

The greatest power that there is, is the power of love directed down from spirit, through our minds, for us to put into action for those who, because of the negative vibrations they have drawn to themselves, cannot help or heal themselves. We create either positive or negative vibrations with our thoughts and actions.

Whatever power you give off to others you get that power back. To communicate with God, as in communion of bread and wine, or in communicating with spirit, the mind needs to be on a positive vibration, so that they can draw to themselves positive spirit thinking minds, for the purpose of help and guidance through physical life. So with that knowledge you can help others. If a person opens up on a lower level, because of negative thinking and action, they draw the lower earth-bound spirits towards themselves. They do not have any great knowledge, but they try and make you believe they do. Jesus said, 'Know them by their fruits, a good tree bares good fruit.' They play around like children doing more harm than good. There is no hell, except that which we create for ourselves by going down a negative pathway. Nobody judges us, we judge ourselves, once we have seen the full concept of all thought, word or actions, in this physical lifetime. Most people have passed onto spirit before they have come to this realisation. It is easier and quicker to learn your lessons, that are necessary for you to advance, on the earth plane. On no other plane does the greatest and the lowest share the same level. Once a lesson is truly learned that lesson will not return in your life. If you do not learn that lesson it will continually keep on repeating itself until that lesson has been learned. Hell that we create for ourselves is a lonely, sad, confused place. It helps to know a little about the spirit world before entering into it.

We all go through a great struggle within, the light of ourselves struggling against the darkness of ourselves. You can only do your best – that is all that is required of you. To those that cause you harm, pray for them, because they are truly lost in their karmic darkness. If people will not listen about a better and truer way, you need to stand back and let them, for a while, go their own way. If you do not they will drag you down to their level, and eventually you will learn when to enter into a situation and when to stay away, for your own good. All that I have written, Jesus told in parables, two thousand years ago.

The spiritualist church of today, although it is still young as a belief system, is looked down on by many, as witchcraft or the work of Satan. All we try and do in our small ways is what our true teacher Jesus taught us. Follow me, and as spiritualists, we do. The Venusians taught the people how to develop spiritual gifts, and in so doing created a better life for all. The people abused their spiritual gifts, creating a darker, more violent world. So the Venusians who gave the people the knowledge took away that knowledge by making it a law not to use it. Do not communicate with spirit. As the earth has greatly advanced since this law was created, it is not necessary in today's world. At the time Jesus passed over into spirit the void was bridged for all time. Jesus took the beliefs of the Jews and brought the laws of the past up to date. He gave life to a near-dead teaching, took the new revived teaching out of the churches and onto the street so all had a chance to be part of the light and be saved from the clutches of darkness and going down the pathway of immorality as many of the priests of the churches had already gone. Does not Jesus say about the priests, 'Do as they say, but not what they do'? You attract to yourself that which is like yourself. For a person who claims to be possessed by spirit, by a dark negative evil spirit, that person must be at that level of evil, to attract that evil. Live a good life, and that goodness will protect you.

Jesus raised the level of consciousness of the possessed person. Raising them to a level where the evil spirit could not go, so any spirit that had tried to take over this person would be dislodged from them because of the higher level that person had risen, through the love and power of Jesus. Possession is very rare, mostly in the mind of weak, mentally sick people or those who play around with a world they have very little knowledge of. All communication with spirit needs to be channelled through a person with a strong mind and serious thought. When we open up to spirit we open a door within our minds, and once opened, the door is never again truly shut. In Jesus's day people thought all sickness came from being possessed. Very few actually were possessed. If a person thinks that a greater power than themselves will heal them no matter what belief system he or she has, then that thought will open up channels of pure energy, pure light from above. Through a mediator, like Jesus, or through many lesser healers that are mostly connected to the spiritualist movement, a belief in any positive thing brings forth that healing power.

We can sit at night time, when all is quiet, ask for the healing, relax and allow our minds to think of nothing. If you do this for half an hour you will sleep better, so awakening up to a healthier tomorrow. Soft music helps to bring about that frame of mind so the healing power can flow. The Bible is built up of many levels just depending on the spiritual level of the writer of that passage you are reading. The Bible was meant for all, whatever level you have achieved. There are a set of straightforward laws of how to live a good and happy life. Then there are symbolic interpretations which will change from person to person depending on their level of understanding. That is why there is so much confusion; we are all on different levels of spiritual advancement. There are many levels of truth – because somebody believes different to you it does not mean their belief system is wrong. They could be on a higher or lower truth than yourself. It still is the truth, whatever level it is perceived upon. That is why one religion should not pull another religion down.

Some of the Teachings of Jesus

Jesus said, 'When you pray, go to a room on your own so only God knows your thoughts.' In this room, relax and meditate on the greatness of God, and slowly quieten your mind. Peaceful and controlled, that is the time we hear spirit voices, a small gentle voice encouraging us onward and forward in life. The mind is like a blackboard, full of thoughts, and we need to clean that blackboard, for in silence there is room in the mind for a spirit voice to be heard. An active, uncontrolled, confused mind cuts us off from the guidance of spirit. It says in the Bible, 'If your eye causes you to sin, pluck it out.' This does not mean to literally pluck your eye out, it means take note of the part of your body that is making you sin. Looking deeply at yourself, find the negative within, then remove it from your life. Turning the other cheek is not advisable in this day and generation. If attacked, use only as much strength against another as is necessary to control a situation. Face all aspects of life with a controlled quiet mind, not an uncontrolled anger. If a person thought about the consequences of their actions before they committed them they would not bring so much grief on themselves and others.

God knows all our wants and desires so the greater prayer is not for ourselves but for other people. Richness in the spirit would depend on the creative ability of the individual. We create our own world by the power of thought. The creative ability would depend on its strengths and weaknesses on the level of advancement they had reached. An earthly poor man could live in a mansion, and a rich man on Earth, a wooden hut spiritually. Rich men or women with healthy bank balances find themselves spiritually penniless. No matter who we are, we all will enter spirit one day. Always try to be in control, be the master of your own life. If anything is trying to control you, money, alcohol, drugs and even people and also religion, if it does not allow you to think for yourselves – get it out of your life, because whatever that thing may be will crush the life force within you and lessen the amount of power of God's love that flows through every individual one of you.

Logic is the key, use the logic within yourselves when listening to the priests and churchgoers. Think for yourself; do not follow the herd. Everything in life, to survive, has to go forward. Most churches are stagnating in their old ways and worn out doctrines, clothing the truth in many unnecessary ceremonies. A priest is like a copying machine; if the original sheet is wrong so will be every other sheet. Trust in God and also your own judgement of life. The truth will come in your quiet moments. Ask and you will be given, knock and the door will be opened to you, in God's time, when he or she thinks you are ready. God is both sexes in one being or power. The Jewish belief is a male-orientated belief so their God is male. Some, like the Egyptians, worship goddesses. If you think of the highest spiritual standard of the male species entwined with the highest level of the female species as one being, one ultimate power, we go back to being part of one power. Jesus had developed all his spiritual senses, healing, clairvoyance, prophesy and astral projection, also a control over the elements of our world. We are spirit, connected to a physical body by a golden cord. At physical death, the cord becomes weak, thinning itself out. Then it breaks, leaving the spirit free to go back to its spiritual home. The cord from spirit connecting the physical, once broken cannot be connected up again – except being born again as a baby. Yes, Jesus was able to heal, but he could not bring any body back to life once the cord was broken. They slept in a coma and were buried this way. In that day they did not have powerful machines to tell when death occurred. Many were buried alive. For healing to take place a person needs to have reached a level where he has earned the right to be healed. What makes a physical body crippled, is many incarnations of lessons not learned, many unproductive negative lives, forming in physical life as body ailments.

'Do you wish to be healed?' Jesus said to a cripple trying to enter a holy pool of water in Jerusalem, which every day was stirred by an angel. Whomever entered the pool first was healed. 'Do you wish to be healed' possibly means, do you wish to break away from the karma of your past, and not to practice that pathway again, that which has crippled you in the first place. Healing is a renewal of spiritual energy to the starved body, and awakening of the mind. A change needs to take place for the better within the sick individual. Jesus says many times in connection with healing, 'sin no more.' Produce negative vibrations in your life, and you will feel the effect. How you are is based on how you lead your life. There are a lot of genuinely sick people in this world, I hope they learn through the silence of the mind to tune into that healing power which is around all of us. But some people surround themselves with so much grief and suffering which does not need to be. They spend their lives moaning and groaning about life, instead of looking for a positive out of the negative. If you have one you must have the other. These people do not wish to be healed – they enjoy being ill. Nobody expects anything from them, a soft life looked after by others. These people achieve nothing, stagnating their own spiritual progress. 'Look at me, how I suffer.' A martyr to their own negative thinking. This is why Jesus said 'Do you wish to be healed?' No two people looking at the same piece of scripture will have the same understanding as another. So I am going to write two different possible understandings of the same scripture, to allow the reader to make up their own minds, to think for themselves.

I refer to the feeding of the five thousand. He had the power to multiply the seven loaves and a few fishes. He had proven his ability many times in the New Testament. It could have happened another way. A miracle within a miracle. The people had been with Jesus all day, everybody was hungry. Some had brought food with them, most of them did not. Why should we, with food, feed those that have none? So they kept the food for themselves. A little boy went up to Jesus and gave him all he had. This made the rest feel guilty and low in spirit for their greed. So they all went forward with all their food that they had hidden away. And so five thousand people were fed with seven baskets of food left over. A lesson well learned for the people that day.

Baptism is the cleansing of the soul, out of the true and sincere desire to change. This awakens the mind to a higher, greater plane of thought. Just depending on how strong your desire is to stay in the light is symbolised within the story Jesus told. Take the parable of the sower, chapter 13 in Matthew. The seed being knowledge of how to truly live our lives according to God. As you have originally come from God, you will return to God, to that spiritual home you have created for yourself in paradise, by your good deeds and positive outlook on Earth. Like the sower, in the Bible, some people start off with all good intentions. But their old negative ways draw them back, choking out of them the little truth they had. Some people understood the truth, changed their lives from the darkness of their negative ways, to the positive, living every day in the glorious light of our father and mother God. For these were people Jesus came down to the earth to save, to save from the darkness of themselves. The world sees people, as people wish to show themselves to the world. Jesus could see in any individual what they are truly like.

He knew in the Garden of Gethsemane that all would scatter; he could see into their very souls. Matthew chapter 8, verse 21 says, 'Lord let us go bury our father.' The light was with them in Jesus, and in the light is life. The spiritually dead live in darkness, so let the dead bury the dead. We grieve only for ourselves, missing the physical touch of a loved one. Then our sorrow turns to anger at life, at God, usually at the person who has just passed on. It is a time of sorting out the memories of the past, of sorting out the good and the bad connected to the spirit that has just parted from you. Your anger builds up as you remember all the bad things they have done. In the end of this period of sorrow the only memories left are those of when you were closest. Thoughts about that spirit person are of love. As a person passes into spirit their true spirit self, their spirit body, is drained of the energy of life that flows from our father God. Along with the grieving loved one on the earth plane, spirit grieves too. All the suffering that spirit had caused in physical life they would be going through with their loved one on the earthly plane, with heightened spiritual senses knowing the wrong, knowing why. This draws the spirit closer to their partner, this is when they try so hard to make contact with their earthly partner. At the worst point of grieving the spirit tries to say, 'I am sorry, I love you, it hurts to see you so sad.' Communication between spirit and physical only happens when a bond of love flows from the person on the physical to the person in spirit, a belief system in a loving God and that life goes on forevermore. We part for a while and we will meet again, bring hope, bringing the grief to a quicker end, if spirit communication takes place.

At the point when love rises from a physical loved one to a spirit, creating a strong bond of love this is like saying to a spirit, 'I am OK, I am slowly building up my life again.' This allows the spirit to go on, to the place where the spirit body can be healed and strengthened, and to be re-educated in spirit life. Through the suffering of others in the physical life, this draws the spirit being close to the earth plane and does not allow them to go home. The greater and longer the grieving of the one on the physical binds the one in spirit to the earth. If you truly love a person who has passed on, you need to let them go for a while, to allow that spiritual loved one to go home. Life goes on, no matter what we go through. Parting makes the meeting again more glorious. If a person on the physical wishes to contact a loved one in spirit, take a flower as a gift, hold that flower up, thinking of loving thoughts about the spirit loved one. Better on a birthday or a special day between yourselves. You should feel a presence in the room, a warm loving presence that you will know without a shadow of a doubt is your loved one. Jesus dying on the cross and rising from the dead on the third day should have proved there is no death, and that life goes on forevermore.

Crucifixion

A dark and lonely pathway lay ahead, and Jesus knew what suffering was to come his way. Willingly he entered Jerusalem, when at any point before that time he could have just disappeared into the crowds and waited until the Passover had finished. Giving his life for many, so we all can be saved. Bridging the great void which up to that point, the day of ascension, saw the human race cut off from their creator. Their one and only true God. The gods of the past were Venusians, planting the seed of light, so Jesus could be. The Earth as whole, because of Jesus, took a mighty step forward on that spiritual evolutionary pathway of light and life. The new age, the golden age, the seventh day, the Lord's day is upon us, opening up around about the year two thousand, which could not have been if Jesus had not willingly accepted that pathway to the cross. A lonely pathway, in the end, at his point of greatest suffering. All that were close to him abandoned him, even Peter. Peter also denied knowing him three times in the court area where Jesus was tried. For what sin they never found, for he was as pure and sin-free as could be. At the last supper with all his friends around him for the last time, Judas, whose purpose in life was to betray Jesus, could not do his job; he loved Jesus and backed off from the job of betraying him. Jesus saw this and said whoever it was who would betray him tonight would put his hand in the meat dish at the same time as himself. Jesus knew it was Judas. The meaning of this act was to tell Judas, 'I have come too far, go do as you should, do the job that you were given physical life to do.'

Judas left and betrayed Jesus for thirty pieces of silver. Judas was doing a job, the rest abandoned him out of fear. Judas through the ages has been painted and written about, showing him to be evil, but in reality he was a good man doing his job, who had a great love for Jesus and more than likely had the greatest love from Jesus. Later on that same night Jesus and his disciples were in the Garden of Gethsemane. In the garden Jesus gave all of himself to God when he spoke the words, 'Not by my will, your will only oh Lord.' He willingly gave up his life, a perfect being, spotless of sin, for you and me. His life and his words have a far greater meaning and understanding today than they did in the time he lived, as we enter a new and more wondrous time. Jesus said, 'Why do you marvel at the things I say or do, do you not know that there will be a time when all shall do as I do and greater things will they do?' It is so wonderful to be on the side of the Lord, the right side for all.

Jesus was whipped, beaten, tried in a court of law, found guilty of no sin, but still crucified. All that he had done seemed to be used against him, and nobody, not even his own beloved God, was with him at this hour of great suffering. At this point of his lowest possible level of suffering and pain he screamed out, 'Lord, Lord why has thou forsaken me?' We have problems in life, and our problems and sufferings are as nothing compared. Then God shone his light and understanding on a thief who was being crucified at the side of Jesus. 'I have sinned and this crucifixion I deserve, but you have no sin, you are innocent.' Jesus saw the light of his father in the words of this thief, and said, 'Before the night is over, you will be with me in paradise'. This lifted Jesus's spirit, enabling it to feel the flow and power of warmth and comfort. Jesus said, 'It is finished,' then he passed on into spirit. A soldier sliced the side of Jesus with his sword, and water poured from the wound. Everything is at a specific level, there are many levels of the same thing. For example ice, water, steam quickening up or slowing down the vibrations in that object, on that level. All liquids are first formed in water, that would be in the first and the last. A pure Jesus, pure liquid at its highest sense running through his veins. I would have been more surprised if it had *not* been water.

Jesus was laid in his tomb, a cave carved out of rock for the purpose of burial. In the Roman courts in Jerusalem the Romans said, 'If he is the Messiah, it is written on the third day he will rise again. More than likely his friends will steal his body and say he has risen. This would give us more problems with controlling the Jews, greater than when Jesus was alive. We will send guards to guard the tomb day and night, at least for three days until the prophesy that he will rise, passes by. Then his life will be like all the rest of the prophets that have come and gone.' So Jesus's tomb was surrounded by guards. God brought over the soldiers a deep sleep, the spirit world moved the stone in front of the entrance to the tomb by channelling power to the rock: spirit being the greater force. On the third day Mary Magdalene found the tomb empty and the stone rolled away.

'Where is my master?' she cried to one who she thought was the gardener, but was Jesus. She did not recognise him as being Jesus. Whether his higher true form is different in some way to the physical form she was used to, I do not know. Jesus made himself known to her. 'Who do you look for?' said Jesus. 'My master,' said Mary. 'I am he,' said Jesus. She went to embrace and kiss him. Jesus said, 'Do not touch me, I have not yet ascended to my Father.' Seeing our spirit and physical are one in physical, parted at death. The suffering of one is reflected in the other, his spirit would have been weak so he would not have the power to form as a solid human being, which he did in the upper room to his apostles. If she had touched him, her hand would have gone through him, for Jesus was spirit. With Jesus being spirit, this proves without a shadow of doubt that life goes on forevermore. Proven by this example of Jesus returning from the so-called dead.

If a human being passing into spirit who was very weak in power could show himself to one person, you would think he would show himself to the one he loves – in Jesus's case, his Mother, Mary. He showed himself first to Mary Magdalene. It is possible Jesus may have had a wife. Mary Magdalene left Jesus and the tomb to find Mary, Jesus's Mother. Peter came to the empty tomb with the loincloth neatly folded at one end of the slab. Peter saw two angels standing in the tomb, who said, 'Tell all he has risen.' Where did the body go? It was not stolen by his friends. The spirit is higher, of a faster vibrationary level than physical life. As long as a physical body is useful to spirit, that body will stay in existence. As in death the body cannot keep hold of spirit because of its decayed state. If a physical body rises to the same spiritual level as the spirit this draws all energy from the physical, turning the physical into dust. With most of us we are not at that level even to our own spirit self, so it takes longer for our bodies to decay because of a slower emptying of the life force out of the now dead physical body. Three days they leave the body out of the ground – this is to release the life force, and also to prevent disease in hot countries.

Jesus appeared many times to different people as a spirit form. The apostles, scared and defenceless with no leader, huddled together in the upper room, the door locked for safety, and Jesus appeared as if from nowhere, as a spirit but with a freshly charged spiritual body from the light of our Father God. There is a substance which is partly spirit, partly physical which can create for a time a seemingly solid spirit. This is called ectoplasm. Jesus, with his newly powered spirit body was able to create this ectoplasm substance. He said, 'You see me yet you do not believe, there will be many who have not seen me yet they still will believe. Touch my side where the sword cut into, touch my hands or feet where the nails have been.' His body seemed solid to them; this was a build-up of ectoplasm, like a physical cloak over a spirit body.

On Ascension Day, Jesus, in spirit form, rose up to heaven so as to cross over the void by the bridge that he had built by the way he led his life, so that we can all cross over into a new and greater spirit world. After this time Jesus was not seen again in the physical world. 'Go to the upper room and wait, the power of spirit will come to you,' was his last instruction to his apostles. On the day of Pentecost the apostles were in the upper room when a great wind appeared like tongues of fire. Their conscious level, because of the build-up of spiritual power, rose then to a greater, higher extent, opening all their psychic senses, giving them the power and understanding to carry on the great work Jesus had instructed them to do. Thus, filled with this power they had to speak and tell everybody about Jesus and their own experiences of being with Jesus. The power of the spirit had given them the power to heal as in the story of the cripple outside the beautiful gate of the temple asking for alms, who walked into the temple by himself. The priests who saw this healing imprisoned Peter and John for the night and told them not to speak of Jesus.

This for the priests was a situation that was spiritually out of their control, as they had thought Jesus would be no more after they had crucified him, but his followers were becoming greater in number daily.

Further on in the Bible it is written that the greater the pain and suffering got, the Romans, through the direction of the priests or suffering created by the priests, only multiplied the people following into this new religion. Their faith was so great in Jesus that no man-made evil could destroy their faith, their belief in the living God and in his son Jesus. What a belief. It is a pity that we have not got that type of belief in today's world. A deep and sincere belief system is what we need today to be able to survive the destruction of humankind against not just itself also against the Mother Earth, causing earthquakes, volcanic eruptions, floods and freak weather conditions. As one family under the one flag as taught by Jesus in brotherhood and sisterhood. We then will survive as one. The law of the new church had to be hard to be able to make a strong enough impact on the masses to survive. Law changes with every age, law covers that which is necessary for that age. As with Old Testament law against the new law of Christianity of that age. In Acts five in the Bible, there was a man named Ananias with his wife Sapphira who sold some land giving half of the proceeds to Peter; Peter found out. It was church law at that time to sell everything and give it to the church community fund, to help all the congregation. Ananias and his wife Sapphira kept part of the money back, like today we keep ourselves back from God. Archaic laws would not work in today's selfish society.

Up until now I have given you a dim view of the future of our society, and so it will be. Out there in our world today are many millions of true faithful Christians separated by small differences in their belief systems. Great sorrow brings out the goodness in people, leading to one belief, one family. We have the capability to survive, will you the reader be among the survivors, or your children? Teach them what is written in the Bible to give them a better understanding of what is about to happen, and then they will survive. As a parent you should have seen to that in their upbringing. After all, it isn't just your own future at stake, it's your children's future too.

Priests became jealous of the apostles, as they, the priests, were losing power and control over the people, so when Peter and many other apostles were teaching and healing in the streets, the priests had them arrested. Those who walk in the light have the grace, guidance and protection of the light. When the apostles were in prison were they forgotten at their darkest hour? No, the Lord in his great mercy sent an angel to the prison to set them free. When doubt and sorrow come your way, turn your face upward and at that very moment help will come to you. To be in the light is to be protected by the light. If God, who is the light, is with you, then who can be against you? Those who are create for themselves a barren road with their fruitless actions. It is hard but we should feel sorry for them, and at least try to remember them in our prayers. As Peter said, 'O Lord, give me the power and strength to speak of you always, even amongst my aggressors.'

The priests sent guards to the prison to fetch Peter and the apostles that were arrested with him, but they were not there – they were teaching in their temples as they had been told not to do by the priests. The priests had them beaten then let them go. Did this silence the apostles? No, it just made them more determined to speak of the life and love for all, of Jesus. All actions of evil men are fruitless. Elsewhere in this situation, Stephen, one truly blessed by God although his ministry was short, did great works. Even the priests that arrested him and had him stoned to death because of his belief in Jesus, looked upon him and saw it was like looking at an angel. Stephen kept faith till death. Saul, part Roman, part Jew, watched the stoning of Stephen and was moved in spirit. Saul, because of his position (partly Roman) was heading for Damascus to hound out Christians and put them to the sword. He was zealous in his work in bringing down this evil band of Christians and wiping out the name of Jesus for all time from the record books. But watching Stephen being stoned stirred some deep lost feeling inside his soul, making him wonder whether his job was as good a job as he first thought. He still hounded the Christians and set out on his way to Damascus. The Christian population was scattered. They met in small groups so as not to be caught and sent to prison. No matter how great the persecution, the word of the Lord was still spoken and great healing work still took place; there was no silencing the one true faith. They just became stronger and multiplied. Many great and powerful magicians and sorcerers of that day, after hearing the apostles' teachings, gave up what they practised and were baptised. One of the great magicians, called Simon, entered into Christianity to see how it worked, to be able to practice for himself and asked one of the apostles, 'How much money do you want for the power you have?' 'Your money will not buy you the power from God. Repent, change your ways for the better,' the apostle replied.

A pure heart is the only method of being able to sense God and his power in our lives. Money decays, including all that is bought with it. The power from God is everlasting and does not decay. Saul, whose heart had been stirred but not enough to stop him from persecuting the Christians, headed towards Damascus with a letter from the high priests to find the Christians in Damascus, and bind them, men, women and children – it did not matter as long as they were Christians. Then, he was to bring them back to Jerusalem for their trial. But on the road a brilliant, blazing light came down from heaven and landed on the ground in front of Saul. Out of the light a voice could be heard. It said, 'Saul, Saul, why do you persecute me?' Saul said, 'Who are you Lord?' The voice said, 'I am Jesus.' After the light had gone, Saul opened his eyes to find out that he was blind. He needed to be led to Damascus. He was three days without sight, and in this time he neither ate nor drank.

Paul

There was a disciple in Damascus who was told in a vision about Saul's conversion on the road to Damascus, and his name was Ananias. He was told in a vision to go to Stright Street to meet Saul and heal him of his blindness. Ananias did not want to heal Saul because of the things that he had heard about him from the travellers from Jerusalem. The voice in the vision, which was the Lord's, said, 'Go to him, he has been chosen by me to carry my name to kings and the sons of Israel.' Even if we were truly repentant of our sins we must also feel the full effect of our actions, and that is why the Lord said to Ananias, 'He has been chosen, I will show him how much one has to suffer for my name.' We need to know both sides of the coin, cause and effect.

When Ananias put his hands on Saul's blind eyes Saul could see, not just things of this world, but his psychic senses were totally opened to the spirit world, to the creative force we call God. He was truly able to see. Most of us see, yet we see very little about what life is all about. Just as the apostles did, Saul went out into the streets to tell all that had happened to him and what he knew of Jesus. It seems once a person has received the power of the Holy Spirit they cannot seem to contain it within themselves. They need to tell others, anybody who will listen. At first he was not accepted in the Christian movement because of how he had treated Christians in his past. But once they saw him and that the spirit shone through him they accepted him into their midst. He stayed there seven days learning more everyday about Jesus. The Jews that were not of the Christian faith heard Saul speaking and plotted to have him killed. Saul found out and escaped from Damascus, being lowered in a basket down the outside wall. Barnabas, John and Saul in Paphos met a Jewish false prophet named Bar Jesus. He tried to stop his master hearing the good news. Saul, now called Paul, saw this Bar Jesus's intentions, and so made him blind for his evil actions. Paul and

Barnabas travelled together, preaching the word of God on the way. Wherever they went, great crowds would come to hear what the apostles had to say.

Many Christian movements still mistrusted Paul. Barnabas, who was travelling with Paul was looked up to in the Christian movement. Paul was accepted on the word of Barnabas. Paul preached so strongly the faith of Christianity, and the Hellenists got to hear of Paul and how Paul made them out to be false, so the Hellenists made plans to kill Paul. So Paul was sent to a place of safety, with Barnabas, in Tarsus. A great but short-lived time of peace followed in the Christian movement, allowing Christians to multiply greatly. There became a great famine over the land so Christians helped other Christians that were a lot worse off than themselves, so that all could survive this time of sorrow. Paul and Barnabas were sent to Antioch to collect goods from Christians for their poorer brothers and sisters. Christianity was growing fast, becoming a strong force to all whom heard the word, and many followed. The Jewish priests and Roman governors, who had ruled for so long found Christianity a threat to their power, and their position amongst the people. So they tried to imprison or put to death anybody who dared speak against them. Paul and Barnabas heard of their threats to stone them so they moved on, more determined in spirit and stronger in faith, in teaching the true word of Jesus our Saviour. In Lystra, a cripple listened to what Paul had to say, and by his faith, Paul healed him. If a person is ready to be healed and has learned the lessons, that made him a cripple in the first place – possibly some karma gained in some past incarnation – then healing can take place. If a sick man practices the same as what caused his illness, and he has no desire to change, then even Jesus, the great healer and teacher, would have problems healing such a person. I doubt if they could be healed. Sickness is the effect of wrong living, so you sort out the cause, and then a person is ready for healing. What is the point of healing a man so he can cripple himself again by wrongdoing?

The people saw the cripple walk and thought that Paul and Barnabas were gods. Paul tried to tell the people they were human, just like themselves. All will be gifted eventually; it is the next spiritual evolutionary step for earthly humankind.

Some Christian Jews were teaching all who follow Christianity should be circumcised, as in the old laws, and the teaching of Christianity should be for Jews only. The Venusian seed was developed and protected in the womb of Israel. Circumcision was a way to signify who was of that race. After Jesus was born there was no reason to keep one true pure Israelite nation, so the law of circumcision is no more. As for who can be a Christian? Peter, while waiting for his dinner to be cooked, had a nap on the flat roof of where he was staying. Something was on his mind: he had to go to Jerusalem to speak to the Christian congregation on this subject. Sleeping in the heat of the afternoon, he had a dream. If we as a people learn to ask we will be guided forward, without falling unnecessarily into sorrow that need not be. Ask your Lord or spirit friend questions just before entering into sleep, and the answer will come. Possibly in dream form – there are many ways spirit influence our physical world – you will get your answer, just as Peter did in dream form on the roof. He dreamt of a pure white cloth tied into a bundle. The cloth was holding something, and slowly floated down from the blue sky and landed at Peter's feet. Peter was hungry in his dream; the cloth opened and every foul disgusting thing that ever crawled, walked or flew lay on the cloth before Peter. 'Eat,' said the Lord. 'I cannot, it is unlawful to eat of unclean things.' 'I am the Lord, who are you to say what is clean or what is unclean. What law was taught yesterday, is not the law of today. Law of the past old and gone. All creatures are now clean. Did not I create them?' said the Lord. 'My son died for all not just the few. Eat.' So Peter did, finding the food far greater in taste than what he had ever eaten.

Many miles away, a Roman centurion, a man of great power with many men under his command, had been told of Peter and of the wisdom he taught. So he sent a messenger to Peter, 'Come stay with me Peter, and tell me of your God. I have heard some and wish to know more.' Peter went straight to the centurion's house with the messenger. He stayed a while and taught the centurion and his family, his servants and his slaves about Jesus. Before Peter left all the family, the centurion, his servants and slaves were baptised. From this day Christianity has been taught to all people, Jew or Gentile.

From there, Paul, with Silas, went to all the places they had preached before, to see what progress they had made. To give your life to Jesus is but the first step. A Christian needs to continually try to find ways of improving themselves so they are always advancing forward spiritually. No matter what a Christian comes up against, the Lord is with him or her. If the Lord be with you, who can be against you?

While Paul was at Athens waiting for other apostles to meet him there he noticed a city full of idols and tried to teach about Jesus to who ever would listen. Some wished to know and asked Paul to tell them more. Paul said to the people of Athens, 'I have seen the things you worship and have noticed an altar to an unknown God.' Paul spoke to them of a God that created all things. 'He is not unknown to those who wish to know him and to know him through the teachings of Jesus is the only way to salvation.' Paul was called by the Christian fellowship to go to Jerusalem, warned by many he met on his journey that in Jerusalem he would be imprisoned and then put to death. Knowing this he carried on his journey to Jerusalem. He said, 'If it's God's wish that I should die in Jerusalem, so be it.' As Jesus had entered into Jerusalem so many years beforehand, only to meet his death, so Paul was about to follow the same pathway as his teacher. The Jews in Jerusalem wished Paul dead and brought against him false accusations like they had with Jesus. Paul was part Roman so only Caesar could judge him. So Paul was sent to Rome passing from one Roman governor to another, all who said that he had committed no crime to be sentenced to death for.

On a ship on the journey to Rome, Paul warned the captain and the ship's owner that there were great storms ahead, that would result in much loss of life and cargo. Nobody listened to Paul. Just when the captain thought he and the ship were safe, a great mighty north-eastern wind powerfully blew their way. The ship was only a few miles away from a safe harbour when the wind blew and took them off course. The sky became so dark because of the storm and neither sun nor stars were seen for many a day. Hope in ever being saved was totally abandoned by all on board, except for Paul and his faith in our Father God. There was no food, and everybody was waiting for that inevitable watery grave. Paul walked towards them and said, 'I told you so, you should have listened to me.' While the storm was at its strongest, Paul said. 'I was without fear, because an angel was standing at my side telling me I would go to Rome and because of this life, cargo will be saved and only the ship will be damaged. Run her dry onto an island.' The crew of the ship were hungry. Paul came up with food and plenty to spare. Is this a smaller version of the feeding of the five thousand by Jesus? Two hundred and seventy-six persons that day were fed, when there had been no food. They soon came across an island. Rocks were hidden under the water, which wrecked the ship. They all swam for shore, but the guards shot the prisoners saying, 'If we let them swim to shore they may escape.' The captain said, 'Do not kill Paul, if it was not for him and his God we would all now be dead.' The island they had escaped to was called Malta. The natives were very friendly, when sitting around the fire, a poisonous viper bit Paul. The natives said, 'This man is a murderer!' Although he had escaped the sea, fate had now come to take him another way. Paul just waved the snake off unhurt. All the natives watched Paul waiting for him to die from the poison. After many hours, Paul was still alive and well, so they changed their minds and now thought he was a god. Paul saw the generosity of the natives, and saw also that some were sick, so he healed them.

Paul eventually reached Rome, and Christians came as close to Paul as much as they dared. They, the Romans, gave Paul his freedom, but only in Rome; he was not allowed to leave Rome. So he preached to all he could, all that would listen whether they were Jew or Gentile. The Earth is a school of learning, as a race we were not learning. Teachers were sent to us from above. The greatest of those teachers was Jesus. Without Jesus teaching in this world, showing a new and better way to live your lives, we would have as a race created a negative physical existence. Bringing about the same fate as that which had befallen Sodom and Gomorrah – they were warned but did not listen. Darkness brings sorrow, pain and destruction. Light, being the teachings, of Jesus brings contentment, happiness and depending on the strength of your faith, protection from darkness within the love of the light. We need to be united as one, in one faith, to fight or protect ourselves from a common enemy as Paul speaks of in Revelations. The sun feeds off power from the planets, eventually engulfing a planet. The sun is active because it burns off fuel, which is hydrogen. When it is low on fuel it draws the planet closest to it, Mercury, slowly towards itself, feeding on the hydrogen from that planet to keep itself well fuelled. This happens every 2,000 years and is happening now. The sun only takes the amount of energy that it needs. Eventually Mercury will become so close to the sun that it will one day become engulfed in the sun. If this does not happen, then the sun, once it has burned off all its fuel, will become hollow. It will then cave in on to itself, creating a supernova. So it has to refuel itself from Mercury in order to keep on being active. Recharging its own power, all planets will sense a stronger sun influence and this stronger influence activates more strongly the Earth's core, which is the same molecular build-up as the sun itself. Power draws the same type of power as itself to itself. With the Earth's core heating up, this will cause volcanic eruptions, earthquakes and melting ice caps leading to floods. It is a natural forward step for the planet's evolutionary pathway, not something the human race has caused.

The human being is limited on how destructive he or she can be. They can destroy themselves and wipe out everything on the surface of the Earth but cannot do much to the planet itself. After the human race has done it worst the Earth has the capability to heal itself. This disaster will happen, and no human being, or God, for that matter, can stop this age from beginning. Jesus teaches brotherhood and sisterhood and to be united as one. If we do not follow Jesus's teachings and live them moment by moment, as a race we will not be ready or prepared for the oncoming disaster. If God is on our side – with God being the light and that light being in all our souls – however great or small the disaster to the human race, the race as a whole will survive. Separated, at war with each other, creating darkness instead of light, we will bring about by our own foolish actions a great sorrow to all human and animal kind. As the body has many parts we all should be part of that one body. It says in the Bible if your enemy is thirsty or is hungry, feed them and give them water. In this positive action towards a negative situation, you will create in your enemy a heightening of his or her conscious level bringing to them great suffering and pain as they realise what they could have been and what they are. There is no greater suffering than when the mind is in conflict with itself.

John

Astral projection is when a person, usually in sleep state, is able to fully feel their spirit loosen itself from the physical and roam around the spirit world, in order to be given instructions for a future time. Paul says, 'I die nightly,' meaning he astral projects nightly. John says, 'I was in the spirit, the spirit is connected to the physical by a golden stretchable cord, so the spirit will always come back to the body.' While in spirit John was told to write down all that he saw and send this information to the seven churches. He turned around to see who was speaking, and it was Jesus. John saw an open door, opened by Jesus, by the way he led his life. A door once opened cannot be shut; open for all to travel through. John was invited through the door. All that God had created, all that is human, animal or bird gave praises to God.

A scroll, containing knowledge of a past time which the human race need for their future. Only Jesus had earned the right to be privy to this great knowledge. The first seal revealed a white horse with a rider, a bow in hand, ready for battle. The horse and its rider represented truth and understanding of the light. A separation of the light of human beings from darkness. Opening the second seal revealed a bright red horse and rider, red like the colour of blood – this signified the battle between good and evil, bringing great suffering on the Earth until the day the light truly shines: a new day, a new age. The third seal showed a black horse and rider, this being the darkness of this Earth, as the white horse and rider is the goodness. Opening the fourth seal, was represented death as being a pale horse and rider. Many will die in the struggle towards the light, but those that have the light within themselves, although they die, will gain life eternal in the spirit world. This is represented in the fifth seal also, as souls waiting to enter heaven, shown with the white garments they were given to wear – the colour white, meaning their acceptance into heaven. The sixth seal shows us what is to come: earthquakes, volcanic eruptions, freak weather changes, islands sinking under water, new land forming and also floods because of the warming of the ice caps at the poles. On top of this, meteors will rain from the skies, as Mercury is drawn into the sun. Many loose rocks will head towards the Earth. After all this is over 144,000 souls that had laboured greatly on the Earth will rise to heaven, followed by all the rest that had been saved because of the light that shone in them. While the destruction by the Mother Earth is going on, darkness in our world will be at its strongest. The Lord has not allowed us to suffer alone without help. Many people from different systems to our own will, in physical bodies, direct us out of the destruction and teach us the new way forward – we know them now as UFOs that travel across our skies. To change a government you need spiritual-minded leaders. These high spiritual beings have been taking on physical bodies and entering onto the Earth for some years now. They will rule the earth or create change in the Earth's negative thinking for the better.

There are 144,000 of these beloved spiritual beings trying to create light in this darkened world, and these will rise firstly to heaven before the rest, and they are known in the Bible as the first fruits of the harvest. If these spirit people had not come down to Earth or if our UFO brothers and sisters do not come to our aid, all human life will be wiped out. In Revelations eight, it states that a third of all that is will be destroyed. Many people's minds will be so affected, that they will scream for death. An extreme form of insanity will cover the minds of human beings on the Earth. This will be all caused by a natural occurrence, the movement of the Earth going forward. It cannot be stopped; but we can prepare ourselves for this time of suffering. Whether the human race is good or bad does not make any difference to the outcome. The Bible teaches us to work together as one family, the only way we will survive as a race. The Lord is a loving, caring Lord, that is why he has sent us people to help and direct us out of this catastrophe. Satan does not exist; we have within ourselves the possibility of great works, or great injustices to our fellow human beings. God or Satan are two different levels of consciousness, within every individual human being. We have been instructed on what will happen, so as to scare the human race into action. We must instil fear into the human race, so they will then prepare themselves. If a Bible reading does not seem logical, then you need to look for a symbolic meaning, which most of Revelations is; earthquakes, volcanic eruptions, freak weather conditions, land that will become sea, the sea becoming land, the melting of ice caps. Meteors shall fall out of the sky, from the breaking up of Mercury, as it heads for the sun.

There has been no situation like this since Noah's day, except this time water will not totally cover the earth. The Lord gave us a sign, as an agreement, that this will not happen again, the sign being the rainbow. In Roman times, as Christianity was first forming into a strong and powerful belief, the Roman leaders saw a way of controlling the people through fear, by interpreting certain parts of the Bible to suit their own thinking. The Bible teaches brotherhood and sisterhood, how to live in peace with one another, within the power of love. The Romans taught of an angry God, of sin and death, of an eternity in a burning hell fire as punishment, and they being the only ones who could save you from sin, for a monetary price. It was a way of keeping control of the masses by the few, through fear. The Venusians came down to Earth to teach the true way of the light. Human beings abused that knowledge, bringing much suffering onto themselves. We are all personally responsible for our actions. Human beings do not like to accept their wrong actions so they created a personality called Satan to blame for their own wrongdoings, a scapegoat: 'I am not responsible for murdering my fellow human beings, Satan influenced me.' When will the human race learn to take responsibility for their own actions instead of blaming a fictitious Satan? The worlds God and Satan represent are the highest and the lowest levels. They are a collection of similar thinking minds, not a specific individual personality. The number of this so-called beast is 666. This represents the seven days which it took God to create the Earth. The seventh day has not yet come, we are in the sixth day. It is the evening of the sixth day, as we come to the close of the Piscean age, entering into the new age of Aquarius. At the time this book is being written, 1998, from the beginning of life to the Lord's day, is seven days. Each day lasting about two thousand years. If you take the start of the Piscean age, at the point Jesus ascended to heaven, you can see where the timescale has come from. The first six means the breaking down of false religious beliefs, the second six, the fall of wrong governmental control, and the third six is a breaking down of our dubious monetary system – that's all these sixes mean. The earth has seven cycles, in any cycle the spiritual development of the human race should be equal to the cycle that they are living in. If any individual is not at the same level as the cycle they live in, the Bible prepares you for the end of any cycle and the start of a new one. Otherwise, they will not be ready to take on the knowledge and understanding of a higher and more spiritual level, that the next stage is entering into. Confusion for these people comes before wisdom, suffering before joy, like a pregnancy and its pain; before the birth of a new age. That birth being the next stage in the human race and the Earth's evolutionary spiritual pathway. Our Lord only tests his or her children as much as they have the strength and capability to cope with it. We will all eventually enter into a paradise state as told in the story of Adam and Eve. Some sooner than others, depending on how quickly they learn the lessons necessary to go forward. Everything is done for our own personal individual advancement towards the light, even our parents were chosen so as to bring the best possible advantage out of the individual person. We need the Bible to guide our pathway, because it is written in the Old Testament that the human race is not fit to put one foot in front of another without the guidance of God.

The beast means the lowest base self of every individual. We all have a part of the beast and a part of God within ourselves. In Jesus's teachings, which were meant to raise the spirit or soul of an individual out of its base level, some people can perceive the highest and lowest in themselves, in the same mind. Religious people draw themselves up from the low to the high (or they should do) as they have the understanding to do so. Others open a state of higher consciousness in themselves through abuse of chemicals, alcohol and drugs.

If religious they know what is happening. If not, they are classed as sick or schizophrenic; two levels of the same personality that can be perceived at the same time. The Bible prepares us for this battle with our lower selves against our higher. For someone who has opened themselves up to these two levels by abuse, with very little Biblical knowledge, it will only create a state of confusion, a form of insanity within themselves.

This age is an age of action through physical accomplishments; the next age will be of the mind. A more psychic and more spiritual age which should have grown and opened up naturally like a flower, coming to full bloom by studying and practising Jesus's teachings. A lot of our youngsters of today, via chemical abuse, have opened their minds to spirit. There are only a few people who can be bothered or want to know what is going on. Spiritual churches should take a greater level of activity in teaching these confused people; otherwise they will end up joining the first new crazy belief that they come across. Some of these people, realising they have a problem, head back to the mainstream religious beliefs. They are so confused they cannot help themselves, never mind others. They have stagnated in time, will not go forward and do not know the answers to the questions of these confused youngsters. These religious beliefs have not got the spirit of God within them. In the days of suffering many churches will fall because the truth is not in them. There are many levels of spiritual consciousness, going from the lowest to the highest. If a person who is opening up to spirit and his or her life is not as the Bible instructs then they will open up to the level which is equivalent to their own spirituality, which could be one of the lower levels. Spirit of these levels think they are wise, they know very little and what they do not know they make up, thus creating a more confused state for the schizophrenic. We need teachers to show them how to rise to a greater conscious state, to learn true knowledge.

Drugs in our young created this situation; doctors give them more drugs instead of just talking to them, which makes them even more confused. This is the insanity of the world as written in Revelations. The ozone layer surrounds our Earth to protect us from the destructive rays of our sun. As Mercury is drawn into the sun, boosting the sun's power, so the influence from the sun to Earth is greater; this is what causes global warming. We cannot do anything about it. We can prevent the ozone layer becoming more damaged by cutting down on certain sprays. These rays from the sun will create a greater build-up of cancer patients. The sun seems good, but it will soon be a killer, if the human race does not stop destroying the ozone layer. To prepare ourselves for the oncoming disaster we need a rescue system working together to fight against this tragedy, which will take a lot of money to set up with the main base possibly being in America. We waste a lot of money on space research. The only planet besides our own that we should be studying is Mars. When Venus became a burner, the Venusians came to Earth, and when our planet becomes a burner, which will happen, we will head for Mars, so we need to study our future home. Other knowledge, although very interesting, cannot help us at all for our future. As for the moon, it is so close to Earth that if Earth becomes a burner so will the moon. Although we will have to endure many natural occurrences some of the disasters are created by human beings, eager for power and wealth, and bringing about poverty, sickness and an early death through disease, for many.

Birth of a Child

Birth and death mean the same. When a person dies in the physical they are born into spirit. We must learn how to communicate with the mind not the mouth. We can travel from one place to another, just by thinking about it. A mind world can create a home if you wish, depending on what level you are allowed to create on, which would be that creative level which you have earned, by your spiritual advancement. Some will be able to create palaces, others old, broken down wooden shacks. The rich are spiritually poor, the poor spiritually rich. Money is not in itself evil, it is a tool, and how we use it or waste it is what matters. You receive that which you have earned and no more or less. We in the physical body are at one level; the level of spirituality below us would to us be hell. The spirituality above us would be heaven. The body decays, the golden cord that holds the spirit to the physical becomes weak and thin. From the beginning of death to the breaking of the cord takes about three hours. Physical life becomes bleary, shadowy, as your conscious mind moves slowly out of the physical and into your spirit mind. When your full consciousness is in your spirit mind you will then, as spirit, hover over your physical. Death is a very pleasant, relaxed situation; the struggle for physical life has gone. Many who have suffered sickness for a long while, find death a beautiful release. The cord becomes thin and then breaks, creating physical death. The spirit which is now you and has always been you even when clothed in the physical, rises from the prone position above the physical and steps onto the floor at the side of the bed, with full consciousness of the room and the feeling and emotions of the loved ones in that room. You are tired, weary, lacking the life force, feeling sad about leaving your loved ones, knowing that soon you will have to. But you can recharge that spiritual battery that passing over into spirit has drained. At first all of your attention is on your family, then as you look around the room you see appearing before you friends and other family that have passed on before you. They have come to take you home.

You begin to rise upwards towards the dark sky, which is becoming lighter the higher you rise. Some float up, some see a door and walk through it; no matter, it still will lead you into a garden: flowers so bright, scent so strong, animals roaming around free, birds singing all at once, their song seemingly flowing together as one. Many more loved ones are there to greet you. Your guide stands in the background watching on. After a while your guide takes you off to your home, showing you the different spiritual activities of souls at work – not work like you do on Earth, boring monotonous work to earn a living but work that for some reason on earth you would have liked to do but never got around to, work that is pleasurable, not a task.

Time is necessary for the spirit body not just to get used to its new environment, but also to strengthen itself by drawing to itself the life-giving force. The spirit also needs to be healed after having left the physical world. There is usually a long time between physical death and a spirit trying to make contact with a physical loved one. It takes power to make contact and on entering spirit, the spirit is drained of power. Alone with your spirit mind, your senses are heightened, knowing what is and the laws that cover the spirit world. Memories of physical life come back clear, like you were reliving it, the good or bad of yourself. This is a time of mental suffering as you see and understand yourself fully. Nobody judges us: we judge ourselves. If memory gets too painful, because of the grief of or own actions in physical life, we can cut off for a while to receive the healing rays. We have to work out, feel sorry, then change that pattern of thinking which caused the negative action within physical life. Memory of a negative form will keep on repeating itself until that lesson has truly been learned. Some poor souls, because of the way they have led their physical lives do not know of the light so cannot ask for guidance of the light. They are lost for a while, roaming the earthly lower planes – some of them not even knowing that they have passed on. These roaming around the places which hold the greatest physical desire, acting as a soul who has lived their life only for the physical pleasures of the bodily senses. A drinker or drug taker or any other strong physical desire will search out people with the same desires.

A drunk spirit would overshadow a drunk in the physical body. Once there is a desire to change from their negative pathway, this sends a positive beam of light or thought towards the level of spirit consciousness above themselves, which then draws towards them a guide who will try to lift that spirit's conscious state so the soul can be directed to a more productive and happier lifestyle. If the guide has instructed the spirit on how to rise and the spirit thinks it is too much of an effort, the spirit will stay where they are until their negative pattern has changed to a more positive and productive outlook, then the guide will come back to instruct again.

We have freedom of choice; nobody can make us rise if the desire is not in us to do so. If we have led a good, productive, physical life and learned the lessons necessary for us to rise to a paradise-like state, possibly with a soul partner, then we will. Eventually, however, your pathway of advancement will be different to your partner's. In staying together you will be creating a stagnating situation for both of you. For a while you will willingly separate from your partner so you can both advance. Even the place you once thought was paradise will become restrictive to you. You will find you cannot learn any more on the level you are on. This sends a light to the next level above you; a guide will then come to you with instructions on how to rise to that next level.

A person can rise in spirit, which takes a long period of time. It is quicker, and the advancement can be greater if a spirit takes physical life. If the guide sees a strong desire to rise in a spirit, they may take that spirit to the next level up, to show them what life could be if they so desired. The spirit would think, 'This is paradise, I wish to live here.' The same spirit believed the plane of thought they were on was paradise at one time. The guide will instruct on the lessons that are necessary to learn to reach this next level, giving the spirit the choice of pathways they could take, one being physical life. Everything is taken into consideration, the lessons that need to be learned, so lessons not learned in physical life need to be learned again. Also working out personal grief over somebody. For example, say the spirit in his last life was a cruel violent father. Taking on physical life again he could be the son of a father who was as violent as he had been. There are always two sides to a coin: on one side, one knowing how it feels to inflict suffering on another, and on the other, how it feels to have suffering inflicted on you. We get away with nothing; sooner or later we will pay in full for all our actions. Parents, friends, brothers and sisters are all chosen to bring out the best within that individual. When the situation has been prepared for physical life the spirit is taken to the garden where he first entered at the time of passing from physical life the last time around. In this garden is a river. This river flows from the highest planes to the lowest, then into physical life. On the day of the spirit's birth into physical life, he says goodbye to all his spirit friends. To them he is about to die, from this level, and be born into a lower level. The spirit enters the water of the river which is warm and comforting, and he is so relaxed he falls into a deep sleep, slowly flowing down the river into physical life. He awakens from sleep to find himself in the womb of his future mother. To bring one child into physical life takes a lot of preparation by many spiritual beings.

Mediumship

Entering into a new age, an age of enlightenment, many of those on the Earth are, in some form or other, delving into spirituality. Many are opening up onto the lower realms, the levels of spiritual consciousness closest to the Earth. Because many in the world today have very little true knowledge and very few teachers, they are creating a world of confusion instead of enlightenment as it should be. Knowledge, and the practice of that knowledge, takes away ignorance and so taking away many unnecessary fears. Once you have knowledge of the unknown, which is all around us if we care or give time to look at the knowledge of anything, the fear goes. No one knows everything; every new step upwards spiritually brings greater knowledge. Old knowledge, but with a greater, deeper understanding, all that is, evolves. As I have said., if knowledge was symbolised as a bottle of milk, all knowledge that we have on this Earth at this moment in time is equivalent to the cream; we have still a lot to learn. Even God, because the light is forever advancing forwards, is still learning. Law is law, and even God works within the law, although the law is a part of God's nature, as we as individuals are also a part of God's nature. We learn all we can of the plane of thought we are on. Eventually we have learned all that is necessary on that level, with still many unanswered questions. Life becomes frustrating as like being in prison and not knowing your way out, so that you lose hope and battle against walls of our society's wrong thinking. A total feeling of being lost, no future, no light at the end of a darkened tunnel of life. Very lonely, thinking you are the only one standing up for good, and everyone seems to be putting you down. If this is how you feel and you have not got any alcohol or drugs in your system to confuse matters worse and if you do not suffer from a nervous complaint or mental illness, then you will need and are ready to take that next step upwards on your individual spiritual ladder.

First comes knowledge of a higher and greater level of consciousness, which by practice you can create within yourself, a desire to rise, and in so doing helping others that you meet on the way. We are all travelling on the same pathway towards the light. We are all on different stages of this pathway. You may wonder what gift you desire, and what you will do with that gift to help others. To rise upwards you have a greater burst of spiritual energy going through your life; an input of power which needs an output in physical life. For example, being a member of a church where you can practice your new-found gift in art or music or writing. There are many gifts – it is our own choice which gift we develop to take the advancement into a higher level seriously. Think about why you wish for this gift and what can you do to help others with it. Know what you want and then go for it. Once a gift has been developed, like a child learning to walk, this gift cannot be taken away. You can raise that gift to greater things or draw yourself to the lower levels by misuse of that gift. Many ask, and things they have asked for come to be. Then they do not want to know; they did not really wish for that which they asked. To learn and to practice any spiritual gift is time-consuming. Our lives are so busy, so make sure you have the time to make that serious commitment to spirit. As with the sower in the Bible, is your commitment to spirit strong like the seed that falls on good fertile ground, or will you become choked by physical life's pleasures to the senses?

A human being travels halfway in the journey they must travel to go forward. Once a person has knowledge of spirit then a guide is sent. The rest of the distance is travelled with a spirit friend. It is not just hard work and determination of the medium but also the work of many spirit people just to work through you as a possible channel. In meditation the strengthening of the communication between spirit and yourself needs to be at a specific time and practised at that time, so spirit can prepare in your development beforehand. They have a life to live too. Test spirit; as Jesus said, 'Know them by their fruits. A good tree bears good fruit.' Spiritual knowledge helps you to know what level you are working on and what level your spirit communicator is on.

The lower spiritual planes know very little. They make up what they do not know. When your light shines in the spirit world it is like an open door, anyone can overshadow, no matter what level they are. To make sure you are working with the best, the highest your mind can expect, live a life as instructed in the Bible, trying to bring out into physical life all that is positive and not negative towards your fellow human beings. Lead a good, positive life, as this attracts higher positive spirit people towards you, giving you knowledge and protection from all that is false. Good attracts good, evil attracts evil. Evil cannot harm that which is good, the power of the light is too strong around a good person. Clean out all that is negative, in thought, word and deed. No one except at the level of Jesus could live a perfect life; *you* are not expected to. To be on that right pathway, to be attempting to clean up your life, to be working with spirit to the best of your ability. You will never be alone again, your spirit friends and guides are but a thought away.

Six students after a night of merriment at the local pub meet back at one of their flats. Three girls, three boys, not one of them with any spiritual knowledge. In their drunken state, they start scaring each other with ghost stories. There are no problems until one young man suggested getting in touch with the Devil, and this with the power of similar thinking minds, on a lower level. One lady was very reluctant to play this foolish game, but went ahead because she did not wish her friends to think she was soft, a cry-baby. A board was brought out with the alphabet and the words 'yes' and 'no' on it. They all sat around a table, with a red scarf over the light that was lit. All other lights were turned off. Everyone placed a finger on the glass, and Earth-bound spirits saw the light of the small group and tried to communicate, coming out at first with what they were picking up from the students' own minds. Earth-bounds are like children, they like to play games. They are at the level that they are on because they know no difference and have no desire to rise. Love strengthens communication with good spirit. Good spirit feeds off love. Evil or earth-bound spirits feed off fear.

Upon seeing the word 'Satan' spelt out and 'I am coming for all your souls, to drag you deep down into the depths of depravity where I live,' the lady, who was reluctant to play, jumped up, breaking the circle and headed for the lights. The earth-bound spirits were enjoying themselves making fools of these ignorant youths. The table rose up from the ground, and flew through the air and, hitting the young lady who was about to turn on the lights, knocked her out cold. The game was stopped, the board was burned. If you talk to any of those students about that night they go quiet or try and change the subject. No real harm was done; the lady who went to turn on the lights had a bump on her head. The young lady had the capabilities of becoming a great and powerful spiritual medium. She went into communication with spirit, as a platform church medium at the back end of her life. Spirit once said to her, 'It is good that you have opened up to spirit, although late on in life. If you had not tried to play with power you had little knowledge of, you would have grown to love spirit and be a good worker for spirit, but because of that night you feared spirit, not allowing any good spirit close. You do good work in old age, but just think of the greater work that would have been done if you had come into spirit communication a lot younger, as should have been.'

Spirit communication is no game; play with fire and you will get seriously burned. Development is a slow but sure process, like a growing relationship with a guide. A build up of trust between the spirit guide and the physical instrument. Look upon a flower as a bulb rooted to the earth, surrounded in darkness, deeply embedded in the soil, as we all are to a certain extent, chasing after Earth's illusions and getting nowhere. Then comes the desire to work with the light. Like a green stem forcing itself upwards out of the darkness. In time we become one with the spirit of ourselves. As like a flower coming to full bloom, reaching out upwards towards the light, slowly like the development within spiritualism. Things that come quickly only stay for a short time, things that develop slowly last for eternity. All that we do is registered in a light around our bodies; the colour of this light is equivalent to the level of spirituality that we have reached. The slower the vibrationary level of our bodies, the darker the lights are that surround us.

This makes us heavy, drawn down towards the earth by the strong desires in ourselves for that which is physical. The physical world is a world of decay, and only that which is spiritual goes on forevermore. The colours in the light around us are brighter, which creates a quickening of the vibrations, allowing us to rise spiritually. Karma is what creates the colour, representing lessons that need to be learned; a plan of this specific incarnation which is brought into being by the good or bad done in past incarnations. We rise out of the darkness into the light, and if we do not learn the lessons given these lessons will keep on coming back into our lives until we have learned. To learn is to advance to happiness and contentment in life. Not to learn brings suffering and pain and a continual reoccurrence of that lesson. We will all rise towards the light, some sooner than others. It is up to the choice of the individual. We are spirit clothed in a physical body for three score years and ten, so the Bible says. The body is a wonderful machine, with us, as a spirit, in charge. We surround ourselves with so many physical problems that we can cut off the influence from our own true selves, wandering through life feeding physically and starving the spiritual inside ourselves. If the physical body is a machine then the frontal mind must work on the same principle as a computer. To get down certain pathways in the computer you need key words. Because our frontal mind is so full of all of life's problems and not giving any space for the influence of your own spirit, then the frontal mind is like a covered blackboard which needs to be wiped clean. The true voices of spirit, our loved ones, come to us in the silence, because it is very hard to keep the mind still amongst the hustle and bustle of physical life.

Just as you can programme a computer, you can buy self-hypnosis tapes, which are obtainable at the White Eagle Publishing Trust, New Lands, Liss, Hampshire, England. To tune into the spirit of yourself, allowing spirit to talk to your spirit, then your spirit must impress that message on the frontal mind. So the first tape would be a healing tape to allow you to break away from the negative, and a lot of unnecessary luggage, and will follow the healing power to refresh and recharge your bodily batteries. The next tape will show you how to tune into the master within. If you do not know how to tune into yourself, how can you possibly tune into spirit? Then there are many teaching tapes on spiritualism. We all need a teacher to follow, and there are no greater teachers of the spirit world than White Eagle, Silver Birch and Zodiac, who are communicating with the physical world in this day and age. This is my opinion only, for I may not have read any greater works thus far, so that's a pleasure I have still to come. There is a great distance between spiritualism and Christianity – too great a gap for the majority of people to come from one to the other. So Zodiac came to our world, intertwining spiritualism and Christianity as one religious belief. Christian spiritualists bridge that gap. White Eagle and Silver Birch are of the National Spiritualist Church. There are many great spirit teachers on this earth; they will come in greater force, giving off the same message to a greater mass of people, the closer we become to the new age, the golden age.

Spirit Communication

Silence of the mind is a state achievable through self-hypnosis tapes. In the silence you create for yourself a place of peace. You will feel safe, secure, warm and relaxed, with regular listening to the tapes. This is your time, no one can hurt you, no worries to think about. All is at peace for a short time with the tape. The place in your mind becomes real. A safe haven, just think about that state of mind and you are in it. Taking into yourself the feeling you felt while in this state. Life getting you down, step out of it for a while into spirit consciousness. In coming to know yourself you will get to know your own weaknesses. If you are scared or timid, why not make your own hypnosis tapes, putting in them a way of strengthening your weaknesses? Once you have listened to one tape you will know how it is done and that it works. I will give you an example: find your favourite relaxation music, and this will be the backing music to your own voice.

This tape that I am going to describe for someone who is timid and scared of situations in life is called 'The cloak'. Go to a quiet place where you are not going to be bothered for about thirty minutes. Sit in a chair or lay down on the bed. If in a chair, sit with legs uncrossed, back straight, head slightly bent forward, looking down, although your eyes are shut. Hands and palms should face upwards, resting on your lap. If in bed, lay on your back, legs not crossed, hands on chest. My example tape begins. Relax, listen only to the music and my voice. You are in your own special time, nothing can harm you, no worries can bother you. You are at peace, warm, relaxed and totally content in this special personal time. Take notice for a few minutes of your breathing. As your body relaxes so your breathing will flow calmly. No rush, nothing to do, just listen to your own breathing. Breathe in the glorious healing power, refreshing and recharging every cell of your body. Breathe out all the negative feelings of life, also all the illness and sickness of body or mind. Let go on the outward breath. Now flowing down from your head to your toes is the healing power which you can feel, like a cool breeze, flowing up and down your body with every breath. It stops from time to time at weak spots, strengthening that weakness, then moving on up the body. So relaxed, so at peace. Your mind still, quiet, patiently listening to any messages from loved ones or guidance from your guide.

We all have a guide; call on him or her whenever life drags you down. With their presence comes the flow of healing power. If you do not call on your guide, at times of suffering, they cannot come – you need to invite them into your life. Think of them as loyal friends who will never let you down. What stressful lives we all lead. Full of problems, frustrations and sorrows. Think for a few minutes of the greatest problem in your life, say it three times boldly. Then three times in your head just as boldly, say it in your head as if you truly mean it: 'PROBLEM, I WILL CONQUER YOU! THEN YOU WILL BE NO MORE!' And then relax, listen to the music, allow it to be a part of you. Give yourself to the music. How relaxing. In this state, the closest to the spirit world as you have ever been, this is the time when two worlds overlap, a time when we can sense their world and them ours.

Sometimes words are spoken usually as the answer to many questions. Listen and eventually you will hear the spirit voices. Not just in this state, but whenever you wish, on all levels. There are many people around us tonight from spirit. Many of your friends and relations that have passed on. They know your problems, they have brought a present for you; a cloak, a pure, white radiant, vibrating cloak. A cloak of protection from spirit to you. They are putting this white cloak of protection on you. When life's problems get you down, sit down, relax, think of the cloak, think of how you feel now. All is at peace again. Most problems are blown out of proportion – silence puts their proportion back. Now go back to the music, listening for anything from the spirit world until the end of the tape, then slowly come back into full consciousness. Once awake, drink a glass of water just to make sure you are fully back. Hypnosis is a science in itself; in America they use tapes to teach people any subject while they are asleep. A young lady was once put into deep regression to see if she could remember past lives. She remembered being a Jewess in York in the twelfth century. She was being persecuted by the English. She gave some facts that were not known at the time, and later on were found to be true, a possible past memory of another lifetime. We are supposed to learn all aspects of physical life: man, woman, rich, poor, sick, healthy, good and bad. A lot to learn in just one lifetime? I do not think so. From a one-celled organism to what we are today has taken billions of lifetimes for each and every individual soul. As long as we wish to advance we will need to be tested on lessons that should have been learned. The school of learning is physical life.

The last tape you will need will be that which opens up to spirit, and is of a journey through nature, developing the use of your five physical senses on a higher level. For example, walk though a wood, smell the trees and flowers, listen to the birds. What colour are they? How do you feel in this wood? Use all the senses. If the desire is strong enough to open up to spirit it will happen. Everybody has some kind of gift, most have no knowledge of it, while the rest do not know how to develop a gift. Regularly practising its use in the journey you will eventually see, feel, smell, taste and hear all that is said on the tape. As for the one sense you find you can use the best, give that sense more effort to strengthen it by use.

Life in Today's World

You are heading towards the biggest disaster in humankind since the days of Noah. You have running at the side of your system a greater, more advanced system, waiting for human beings to ask for their help so they can guide us safely through this oncoming disaster which they have already experienced in their past. If they were to make themselves known to us, we would blow them out of the sky as we are still animal. We need to develop intelligence spiritually as well as physically. UFOs, as we call them, are bound to be, because of their level of understanding, a peaceful, God-fearing, warless race. Yet, we still would blow them out of the sky or dissect them in our laboratories, as we did in 1947 in America. They already know how they will be treated, and that is probably why UFO's do not come too close to human beings.

In the Bible it is written that God will cut the time of suffering of the human race. If he did not, no one would survive. How is God going to cut this time short? By sending us a higher intelligence, to guide us through. Many do not think UFOs, the higher intelligence, exist; they are our only hope at the moment, and yet we do not believe in them. It has been written that UFOs are the power behind the manifestation of the Virgin Mary. A laser image of the idea of the Virgin Mary, as seen by a young, poor Catholic girl, was the result of higher intelligence picking up thoughts from a child, and turning that image toward her. It is the easiest way to make contact, in creating a channel. This was backed up by good, strong Catholic belief, as France is a Catholic country. That is what creates the miracles – a belief that something will happen. If the belief is strong enough, healing will automatically take place. We have been visited all the way through our history, leaving directions for the human race wherever they appear. When a guide from the spirit world introduces him or herself they may not be what they say they are. They appear in whatever form is easiest to work through the channel they are guiding. For example, Tom, the dustbin man, may have reached a high level of spirituality, but who is the channel going to believe they are guided by – a dustbin man, or a Native American chief? You are more likely to listen to the spirit of a Native American then a dustbin man. The spirit has to have reached that level anyway to be able to guide another. If this is not so, then the only wise people are Chinese and the Native Americans, as everyone seems to have one or the other as a guide.

In our world today many are opening up to spirit, but society labels them mentally sick and instead of giving them knowledge to develop properly, they give them drugs, which makes them worse. Spiritually we have that right, to live our lives correctly or wrongly, bringing ourselves sorrow or joy. Freedom of choice in an everyday physical world depends on the freedom of the society we live in. A poor man in a Third World country has very little freedom of choice. But we all have *some* freedom of choice whatever society we live in, and no matter how small, we should use that right to the best of our ability, to its highest level, the level on which we are allowed to in the society we live in.

We are governed by spiritual law; a pre-set plan of our life is registered in our karma. There are lessons to learn, lessons not learned in a past incarnation, as a light which surrounds the outside of our bodies. If we do wrong the colour of the aura or karmatic plan becomes darker. If, for instance, a man murders, even though he has not been imprisoned by physical law, he has not got away with it – you get away with nothing. His karma will come back onto him, for everything has two sides, like a coin. This man would have the guilt of the murder with him awake or asleep. Plus it is more than likely he will one day get murdered himself. You do not just feel what it is like to murder, you need also to feel how the victim felt, so that man will receive in turn the actions he has done to others. Only when you know the full consequences of your actions can you be truly sorry for them. Any action of a negative nature, in physical life, should be worked out in physical life. The death sentence should not be. As a society we should teach a better way, give them some kind of work rather than send them into the spirit world to be dealt with.

A person entering spirit without working out his or her own

karma in physical life will darken the plane of thought they are on. We are supposed to brighten the spirit world on our return not darken it. To put a person to death shows that the society that sentenced him cannot cope, is not working. The person is more than likely to be sent back into physical life to work out karma. Physical life is a privilege, the only plane we can be truly tested on. If we cut physical life short, in suicides, then we have not run the full race, and the home we would have been able to achieve, we cannot reach. We have not learned enough to be able to live on that level. We must take a lower level. What a waste. No matter how hard life becomes you will think it was easy compared to the suffering you'll create for yourself taking your own physical life. Many people go through so much suffering at the end of physical life as the body is beyond holding onto the spirit, but death does not come. Do we have the right with the dying person's permission to bring a quick end to their lives? We are spirit, and the body is like clothes for the spirit. If those clothes are worn, like an old body, should we not be allowed to disconnect the spirit from the body and allow the spirit to go free? It seems the humane thing to do, to bring physical life to an end. We are on Earth to learn. Possibly the person who is suffering has not fully worked out the karma or lessons needed to be learned on the Earth plane. Should we interfere with someone else's karma? Maybe the person is not ready for death, brought up in a Christian belief, judgement and eternal hell fire for the evil, which makes them subconsciously hold onto life, because of fear of judgement for actions done in physical life. A good person would be glad of death, a release from pain. No conscience-pricking problems with judgement. Then the golden cord starts to get thinner, and they accept death; they do not fight it, so they die quickly and peacefully. If a person wishes to die and can make that choice themselves, they should be allowed to. Does it not say in the Bible that God is a merciful God? There are many levels of consciousness. We go to the level we have earned by our good or bad actions in physical life. Heaven, or hell, as in a home on a higher or lower level, is of our own making.

Karma restricts, so the level will create restrictions. The lesser the karma, the lesser the restrictions, and thus we gain more freedom – a happier and more content existence. There is no hell fire – as an iron rod is hardened in fire, we are hardened against wrong-doing, slowly breaking away from our own karma and eventually rising to that next conscious level. We are not judged once we have reached the level we have earned. That level will heighten our senses so we will be able to perceive a truer version of ourselves. We judge ourselves, learn from our own mistakes, accept what will happen in working out karma. You can go forward, stagnate or go backwards. Stagnating or going backwards brings great suffering. Nobody forces you to do this or that, it is up to you. There is no time limit to get from one level to another. Time is an earthly aspect. There is a form of time in spirit, different from our system. No one can tell the future, for physical life, in the perception of itself, is like a man in a valley surrounded by rock cliffs; the cliffs stretch as far as he can see. Spirit sees physical life as on the top of a mountain. The view and under-standing is greater because spirit is not surrounded by materialistic desires that cut off a physical person's view of life. Because spirit works with reality, we on Earth live in a world of illusion, guided by illusion. So we blind ourselves, not knowing what is true or false.

The law of life is cause and effect, we do that cause but through our blindness we very rarely see the effect until it is upon us. Spirit can see the aura, the light around our bodies and our karmatic plan within the light of this specific lifetime, our past, our lessons not learned. Spirit sees the effect of which we have already done. To say anything will happen with certainty, except death which is inevitable, without studying the cause and effect of that individual, would only be a calculated guess. We travel on one road to the light. There are many roads leading off that road; which road we take depends on our own freedom of choice. So spirit gives off strong possibilities, but whatever road we travel is up to us. What is past has gone, only to be learned from – we cannot change anything. The future has not yet happened. Do not look for the future in the stars for a better tomorrow. Change today, so a better tomorrow will be. Because the present is all that is important. The one true road has many side roads, sometimes because of lessons needing to be learned.

You are allowed to wander for a while, for no matter how far you roam you will always eventually have to return to that original pathway. Whatever you are meant to be, as by your own personal karmatic plan, you will be. The time difference is in how quickly you learn. Life goes on forever, you will reach your personal goal, whether this lifetime or the next, is up to you. The true pathway to the light is long and hard, but you only have to go halfway on your own; the rest of the way you travel with the guidance and protection from your guides and loved ones in the spirit world. In this world today we are discovering many new things. Some for the advancement of human life, for the better, and others like interbreeding, different species, just to see how they will come out just because we can and for no other reason that I can see.

In Greek mythology, it is written about Venusian gods in human form. Pegasus, half man, half horse, is animal and human experiments of today seem to be trying to go as far as the Venusian experiments seem to have gone in creating Pegasus, in meddling in the field of genetics between two different species, creating deformed and suffering individuals, playing God with the lives of our younger brothers and sisters in the animal world. People need to relate to a physical image or symbol, to direct their thoughts to God. When the Venusians broke away from earthlings they carried on worshipping the caricatures of the gods in the things they could see around them, such as spirits of the forest or animals. Statues of these spirit forms, until the image, has greater meaning than the gods they were first formed from. The original understanding was lost. Nothing has power except the power from God. The only power any object has is the power we give to that object by our false belief in it. The power of the cross or any other symbol in religion should be only used to channel thoughts to God, just as one looks at a picture to know or remember a person or situation. So by looking at the cross we remember Jesus and his life.

As I sit in silence, drawing towards myself the glowing, golden healing life force from above, I think of true prayer, not things for ourselves, but asking for others. There is so much suffering, so many lost, sick souls, whether they be adult, children or animals.

There are leaders that are blind from reality, feeding their egos and their bank balances, instead of the two thirds of the world who are starving. There is enough food to feed everyone on earth until their bellies are full. Many thousands of tons of food are wasted daily. Our church leaders are lost down wrong pathways, the people follow like sheep, used to being told what to think so that they now cannot think for themselves. Any sense that has been given – and thinking is a sense – with lack of use will become dull. So a person would find it hard to think logically for themselves and be dependent on a society or priesthood to show them the way. The way is lost to these people, and to follow such as these is going down a path of spiritual destruction. Knowledge is all around us if only we could see, open our minds up towards how life really is, rather than simply be told by those who think they know. The pathway is a personal, individual road. Look for yourself, you will find for yourself.

Help is Out There

There are many levels of spiritual consciousness. The Christ level is the level just below the void. The bridging of the void, by showing the way for all, was Jesus's purpose on earth. We had no connection with the true God, the almighty power behind our system. God's influence stretches from Mercury to Pluto, and there may be planets past Pluto. The only gods we know are those just above us, in our case the Venusians, and their gods would be from Mercury. We will one day be the gods on Mars, so every conscious state or level has a god governing that level, with a greater god above them. On Ascension Day when Jesus rose up to heaven he said beforehand to his disciples, 'Where I am going, no one has gone before me. I am going to prepare a place for you and all who follow you.' As the Christ level is just below the void, Venusian intelligence is the first level across the void. They came down to our earth to draw us up. If the void had not been bridged, life on Earth would have stagnated, slowly bringing an end to life on Earth. The purpose in life is advancement; to stagnate is to lose purpose for existence. No purpose, no life. At this Christ conscious level, many were sent onto the Earth with basically the same teaching: Mohamed, Krishna and Jesus all entered physical life at different times, all from the same conscious level below the void. When a higher spirit takes on physical life, many spirit people are born in that area from close to that level spiritually, to protect and help in the task that needs to be done. If a few great teachers do not create the impact on the masses as planned by spirit, the more teachers there will be, and the greater the chances of the truth being accepted by the masses are.

In Greek mythology they gave their gods, which were Venusians, attributes of men or women, thus losing out on their true nature of being pure bright within the light, knowledge, and powerful, spiritual force. They lowered their gods down to their own standards, thus losing the truth that they brought.

All power has a central point. Take the highest physical power in our system, the sun. Everything that is physical has a spiritual counterpart. The spiritual is the true being, the physical clothing of that spirit. Where is the central point of spirit? The sun. So the sun is our God, from whom all power flows, physical or spiritual. We are but one, in an ocean of many systems. Look through a telescope and we will see how small we are compared to the rest of the universe. We think we are intelligent human beings, but compared to other systems we are backward, still at war with our brothers and sisters in other countries. Beings from other worlds visit us regularly; they are of a greater intelligence and war is long ago in their past. We know only three systems, but most people only believe in one, the physical. The three are the spiritual, the physical and the race that live beyond the vortex, which we call the Bermuda Triangle. Look out into a star-filled sky – every one of them a sun or a planet. So how many must have created a life form of some description, similar to ours? The planets are being drawn to the sun, to recharge the sun, like in our system.

First is gas, which when activated by a sun, becomes a solid. In reality nothing is solid, not even ourselves. We are built up of neutrons, protons and electrons, and so is everything else around us. We are energy that is vibrating at a certain speed. The spirit world is vibrating at a faster speed beyond our perception. To spirit, spirit is solid. The next system to yours beyond the vortex is of a faster vibratory level than our system. That is why we cannot see it. Our sun is drawn to a greater sun and will one day be engulfed within that sun, which is called Vegas. Maybe there is a power link between all systems, a bit like a spider's web and where two threads cross there is a planet or a sun, joining our small system to a greater system, continually unfolding out across a great vast universe, as the Mother Earth's core becomes hotter with the greater power from the sun, and the sun engulfs Mercury within itself. When that day comes we will need to learn to ask and accept help from our higher brothers and sisters in other systems close to us. If we do not, our system will not on its own be able to cope with the destruction that is coming our way. Then a gateway will open up in space, and many races in a multitude of different forms will come visiting Earth, in peace with only one reason, to help and teach their younger brothers and sisters creating in time, not just a lighting up of the universe but a vast family from all parts of the galaxy, working together as one unit.

We are stretching further out into a still-dark universe, pushing out the boundaries of the light. It is a never-ending adventure always moving forward. Our system works only because it has been tried and tested to a number greater than humans can think. We as individuals or nations or a planet need not suffer alone. There is a multitude of helping hands from the spirit world or our UFO brothers and sisters, all watching to see how we develop. Whatever dangers we are about to have in our future, they have already gone down that pathway and survived. Your future is their past, like during the Second World War when the peoples' backs were against the wall. War and destruction of property comes to an end. All the earth, no matter what colour, creed or nationality a person may be, joined together as one nation, under the one flag. One government fighting the same enemy, the destruction caused by the Mother Earth, for the purpose of survival. Only then will human beings ask for help from other systems.

It is written in the Bible that it is not fit for the human race to put one foot in front of the other without the guidance of their gods. Any higher intelligence will have also developed a spiritual level of understanding equal to their intelligence. A peace-loving, caring race, a race without war. A race evolved above all that is negative, that works with universal law, instead of against it, like human beings do at this present time. If we were able to get close to a UFO today, which is unlikely because of their greater intelligence, we would blow them out of the sky. We are still a war-loving world. In 1947, in Mexico, the only reason we were able to capture a UFO was because their craft developed a problem with their cooling system while entering into our system. It is said, and I must confess here I do not know how true this is, that if the UFO had not landed on our planet, we would not have the technology, without using the knowledge gained by taking the craft to pieces to find out how it worked, to have put a man on the moon in 1969. Human beings do not seem to learn much. The UFOs are our gods, and yes, we would blow them out of the sky if we could, just as we tried to destroy them in the story in the Bible many thousands of years ago at the Tower of Babel – we do not seem to learn.

New Earthly Race

Since time began, Venusians have been impregnating our women, to produce a greater and a higher intelligence within the earthly human race. Abraham, the Father of the Israelites, was he the father to his son, with Sara being in old age, or did Sara become impregnated by a Venusian? Same as Elizabeth, who in old age gave birth to John the Baptist. Mary, the Mother of Jesus was impregnated by Gabriel, a Venusian. Whenever, within the history of the human race, that race is stagnating or going backwards, the Venusians have injected our race with Venusian seed and knowledge to return us to forward advancement. We use that knowledge always for negative purposes: once we have suffered for a while then we seem to use some of that knowledge for the good of mankind, that is, the reason we got that knowledge in the first place.

We are again, in this world today, stagnating into a place of greed, power over others, ignorance of true spiritual law and bringing for many sorrow, pain and hardship. Again, the Venusians are impregnating our women to take the intelligence level of humankind one step further forward. The newspapers are full of UFO abduction, which are women who are being impregnated with the Venusian seed. Many experiments are being done on human beings – a little like human experiments on animals. Those who the Venusians experiment on do not seem to suffer any pain. If anything, it's a pleasant experience, with no after-effects. Most of these people report a sense of overwhelming peace. Their lives seem to change always for the better, and they usually move towards religion or a caring profession, which they had no interest in at all beforehand. It is a pity we as humans do not have the same care and attention when we experiment on animals, making sure they are not suffering.

If the UFOs were warlike, with their technology, we would not have a hope in hell's chance of survival against them. The speeds their crafts can reach enable them to be in front of us one moment, nowhere to be seen the next. They seem to be playing with our most advanced and highly experienced pilots and planes. It is a good job they are friendly. Even when they impregnate our women with the Venusian seed, no harm comes to these women. All that occurs is the birth of a highly intelligent child. These children may be the leaders of tomorrow, the ones that will take us all into the next stage of our evolutionary pathway. It mentions in the Bible that 144,000 souls will enter into our world and take us forward into the next age. If you were to put a high spiritual being from the spirit world into a body of flesh, that body would have to be advanced enough to be able to accept that spirit. With Venusian help, 144,000 souls can enter into our world, and possibly prepare our world for the disaster that our future is going to bring.

We are at the back end of the Piscean cycle, entering into the cycle of Aquarius. This inter-breeding may take place at the end of any cycle, to take that cycle forward to the next one. It is amazing to know that children whose mothers have been abducted by UFOs have grown up to be great artists, musicians and great leaders of the people. Such as Moses in the Bible, whose father is never mentioned. In Genesis it says, 'The Gods came down from the skies, looked upon earthly woman, found her attractive and took her for their mate.'

It seems they are still doing the same thing today and for the same reason – to help the human race go forward. It would be a great honour for a woman to bear such a child. With Jesus's father being a Venusian God, Jesus was truly a son of God.

Karma and Law

Like any group of molecules or situations, acts that continually repeat themselves become law. Take water for example. Freeze water and you get ice, heat water you get steam, every time. The first law being that the greater feeds the weaker with power, develops the weaker, then draws the weaker in to recharge itself; the law of cause and effect, and everything in life goes by this law. Example: if a man hits a man, the first man created a cause; when that man hits him back, that is effect. You cannot get by without feeling the effect from any cause. You can not get away with anything. A man in his youth robbed people for his living. He gave up his wrong actions, took a job, and became rich and powerful. Going home one night he found his house had been broken into. All his beloved possessions were either stolen or smashed. 'Why me?' he said. Did not he steal when he was young? Whatever you sow you will eventually reap, whether it be in this world or the next. If we could understand effect, we would produce only causes that were productive to us. What we are, is what we attract towards ourselves, in thought, word or deed. To know the law, to work with the law, brings happiness, contentment and an abundance of what we desire in life. To work against the law brings sorrow and pain. One should be honest, positive in all actions, to ourselves and others around us. How we think is what we are, also what we will attract towards ourselves. All life first formed as a thought, and good positive thoughts attract positive things towards us. Law cannot be got away with, bent or bribed. All you do is registered in light around your body. That action of law is karma.

I once saw a film about a man who sold his soul to the Devil for eternal life. The film was *The Picture of Dorian Gray*. There is no Devil, so he cannot buy a soul. He had a picture of himself, and with every good deed the picture changed to a better, more handsome likeness. With every bad deed the picture became ugly and vile. If you think of your aura, which is the light around your body, becoming brighter or darker depending on what thought, word or deed you have committed, the brightness rises upwards and is paid back in full, and so is every dark deed. Below that layer of light, is a body aura, depending on how you have looked after yourself, or abused yourself. This will show in colour, closest to your body. If a person could see the aura of another (as some can), they would be able to see illness before it showed in the physical body. Break the law and feel the karma, as it comes back to you. Our bodies vibrate at a certain speed. The darker the aura the slower the vibration. Darkness attracts darkness, keeping the body close to the Earth. When you depart from this world into spirit you go to the place you have earned. If your aura is dark your spirit will be drawn towards the Earth, as a lost soul wandering on the earth plane. Your perception of all that is around you is equal to the colour of your aura, so that a dark aura equals a dark lonely, shady world. You may try to contact a loved one – they do not hear or see you. You are truly lost. To rise from this place of sorrow is to change how you are thinking. Be sorry for your actions and ask those above, or God or any higher being for help, and it will be given you.

A guide will come to talk to you and show you how to change your pattern of thought, so you are able to rise upwards towards the light and leave behind the darkness which you have created for yourself by the way you have lived physical life. Your conscious level is heightened so that you see yourself as you truly are. All you have on that level is your own tormented mind. Some do not know of the light, so they cannot call for help or guidance from the light. If your pattern of thought changes, for the better, this sends up a thin stream of light to a higher level, then a guide will come your way. In your wanderings you may see a light on the Earth, a spiritualist church. You may be lucky and a loved one is in the congregation. Everything in life, whether it be in the physical or in the spiritual, has to be earned. If while you have been in spirit your thought patterns are the same, and then you have made no effort to change, whether a loved one is there or not, the guide of the medium would not allow you to communicate. You need to earn that right. The point of a guide is to protect the medium, and allowing a lower form to come close would be detrimental to the medium's welfare. A person or soul, if not ready for communication, will influence and overshadow the medium, with their last thoughts on Earth, which will be the thought they had when they were passing over. The only light that this soul can go near is an open door with no guide as its doorkeeper. If a group of young people used a board game as a laugh, a joke, usually under the influence of drink, there is nothing to stop an earthbound spirit communicating. At first he would try to communicate properly because he has a need to communicate. Once he saw they were playing about he would become frustrated, angry even. Thus, the spirit ended up scaring the youths.

Do not play with spirit, it is not a game. It can be very dangerous. Especially to young, impressionable minds. If a man in the physical world was truly evil, opening a door to him after he had passed over is deadly. If a man came to your door at midnight would you open that door to him, a stranger? Example: a true story, husband and wife, playing with the board for a laugh. They are a couple deeply in love, and the husband would not harm his wife. A murderer comes through from spirit who killed his own wife when he was in the physical, with a broken bottle. The murderer strongly overshadowed the man playing the board game. He jumped up, broke a bottle and rammed it into his own beautiful wife's face. She needed twenty-two stitches. Anybody who has an interest in spiritualism must ask for a guide, a protector. This is the first thing that should be done. Always pray for protection before an attempt at communication with spirit, and thank spirit after communication is ended. If you thank somebody they are willing to come back again, another time.

Average Person Entering Spirit

After the cord has broken and the spirit is released from the body, loved ones are waiting there to take you home. Most people do not wish to go, and they hang around those loved ones left behind in physical life. When they do go they enter into a garden and meet all the loved ones that have passed on beforehand. Nobody has led a perfect life, except possibly Jesus. We all have to work out the negative we have produced for ourselves while in physical life. The difference is that the average person has friends and loved ones to help him adjust not just to the true realisation of himself but to the surroundings he finds himself in. An evil person is alone, with only his negative thoughts as company. We all enter spirit drained of energy from the illness that created the break in the cord. An average person is able to receive that healing, but the recharging of life force may not occur straight away. That would depend on how good or bad he had been. An evil person's home is formed out of darkness, cut off from the light. Cut off from the wonderful healing rays of the light, the evil person is left to wander aimlessly in the dark, with their memories of evil deeds your only companion.

Sickness of the Earth Itself

A person draws towards themselves positive or negative energies depending on how great the light is that shines into their life. They suffer greatly through negative energies building up around their bodies, cutting them off from the light. So does the Earth, scarred by man's inhuman actions towards his fellow man. War, sickness and poverty – they do not need to be. Greed of the few brings about the starving to death of many. The Earth becomes further scarred with the cry of lost, lonely spirits, taken suddenly into spirit by war. A negative energy builds up in these areas, and the land is not productive, stained by the blood of so many. A medium can tune into these areas and see again the horrors that have taken place there. We are all sensitive in some form. The greater your sensitivity the stronger you are able to tune into the past of any land that has had a war on it – cold, damp, chilly places.

If a medium blesses that place in prayer, and a priest also, the light shines, releasing the images of the suffering souls to break up into the atmosphere. The true human spiritual soul would have gone higher, leaving a negative image, like a negative play playing over and over again for anybody who can truly see with the spirit eye. Some mistake these images for ghosts, but the spirit has long gone. Buildings that stand on ley lines, energies that flow from north to south and from east to west – when two lines of power cross we have built churches, pyramids, all holy places on these lines, places of great power, such as the pyramids or Stonehenge in England. When Mercury is drawn to the sun's power, giving a greater influence of power towards the Earth, this will also bring spiritual brightness or awareness to the Earth and everything upon it. Although we all have our own personal pathways to take, building up the light within ourselves, to shine a little bit stronger for the whole, we advance forward, as a whole, created by the light in all life, by the good we do. Everything that is physical has a spiritual counterpart. The greatest physical power within our system is the sun. So God must be the spiritual counterpart of the sun, the power behind the sun. As the sun is classed as a masculine influence, so the moon is classed as feminine. Some worshipped the stars, some worshipped the nature of the Earth. Diana, the name given to the moon, was probably a Venusian goddess. All Venusian knowledge is good, and the human race in its own folly, creates the negative out of the positive. Although true Diana worship is a wonderful belief system seeing that the human race has wrongly used any Venusian understanding, anything in life is like a coin which has two sides. The darker side to Diana worship ruled the people with a mighty hand. For example, offering human sacrifices to Diana for a good harvest and many boy children. For this, a virgin had to be sacrificed on the fire, to purify their race, so they believed. This ceremonial virgin sacrifice, which is so vile and barbaric, is still practised today, on bonfire night, November 5. There has not been a year that has gone by without the death by fire or the scarring for life by fire of these innocent children. Sacrifices are still being made today, but we like to call them accidents. Do not practise evil festivals, even when you have changed the reason or name for that festival. You are still practising the darker side of Diana's religious belief. As long as you practise, the more innocent children will die. Covering up vile, dark, pagan festivals under Christian holy days.

When Christianity went to Rome, in Peter and Paul's day, a great multitude of pagans went towards Christianity. Paganism means the worship of many gods, just as Christianity means the worship of one God. As times became hard they jumped from one religion to another. If Christians were being thrown to the lions, some went back to pagan ways. There was safety in following society's beliefs, which changed with every new leader. So pagan and Christian holy days are on the same days. Practiced positively and not negatively there is no harm in them, except bonfire night. The moon is not evil; it is portrayed to be evil because it shines in the night. Fear is what our minds create when we cannot see or understand a situation. Without the moon's influence we would not have a water cycle. There was a time in

Noah's day when there was no moon. No water on the surface of the Earth, only water in the ground. Everything evolves, even the planet Earth itself. The greater the influence of the sun, the greater the activity in the Earth's core, creating a strong powerful build-up of water under the Earth's surface. Anything that has been once created and has the ability to reproduce itself, will continually create just itself, that is all it knows to create. Building up such a pressure, forcing itself upwards and onto the earth's surface, at the Earth's crusts' weakest points.

Part of the Earth's crust, at its weakest point, broke away and became our moon. There is an ocean on the Earth that is so deep that it is probably where the moon could have broken away from. Rock samples brought back in 1969 are the same as on Earth. When the Earth had reached a point of advancement where it needed a moon, a moon came into being. If a thing has a purpose – it will be. Lose that purpose and that thing will be no more. We can live without a moon as the people before Noah did. But life was hard, and we need a moon to create the type of lifestyle that we live in. Law on the higher spiritual side has always provided for anything that we may need. On the physical level, if we shared, instead of living a greedy, selfish life, there would be enough of everything for all.

With water on the surface, the rain cycle followed. The moon was formed for this purpose. We are influenced and drawn into the sun. One day we will be a part of the sun. The moon is influenced by the Earth, so one day will be one with the Earth. We go around the sun; to us the sun is the greater influence. The moon goes around the Earth, Earth being the moon's greatest influence. Power feeds off power that is weaker than itself. While there is a rain cycle there will be a purpose for the moon. As Mercury is becoming one with the sun, the stronger flow of energy, spiritual as well as a physical power coming, to the Earth from the sun, will heat up the planet and draw the Earth a little closer to itself. As the planet becomes hotter, water on the surface will be no more. Water will go back underground. By day the surface will be too hot for water to stay on the surface. Night, which is a little cooler, like in the desert, will be the only time it rains.

Water collected at night, flowing into water containers just underneath the ground, will supply the cities whose inhabitants cannot walk on the Earth's surface during the day because the sun will be far too hot. Their cities are connected; you need never go on the surface. Here will be a slower pace of life, with machines to do all the work. A time to ponder on greater things. Sex as we know it will one day cease to be. Slowly our genes are being interwoven, the masculine with the feminine, creating a self-productive organ – no need for male or female. There will be a new species formed out of two. Sex in nature will be a meeting of minds, one true spirit mind with another. Could it be better than what we have got? Possibly. There are places you can go, like a glass solar-powered temple, channelling toward a specific spot as it is able to draw spirit friends towards you, then form in front of you seeming solid, so that you can hug and kiss, feel warm to the touch. We will eventually lose the ability to communicate with speech. Conversation will take place mind to mind, as it does in the spirit world.

This planet Earth will eventually get so hot that human beings will not be able to survive on it. We could go to the moon, but only for a short time as the moon will one day be drawn into the Earth, now a burning star. Some will go through the vortex which our Venusian forefathers did. On Mars life has evolved to humans two steps from the ape. The Venusians were our gods, now we will be gods over our younger brothers and sisters, the Martians. If a system is to be found to work that system, it will keep on producing itself until a greater system is found. At the moment, Mercury and Venus have surpassed the rest of the planets, right up to Pluto. Who knows what is beyond? Human life has formed on the third planet in this system. The planets of the past have had human life on them. When the Earth becomes a burner, Mars will be ready with early human life. Mercury will have been absorbed into the sun, making Mars the third planet in our system. Mars has had water on the surface but the conditions were not correct for water to stay there. One- to two-celled organisms will have formed in the water underground, the first forms of life. Channels marked into the surface of Mars, where water once flowed, showed up on a picture brought back from

Mars, by a camera attached to an unmanned spacecraft. The influence of the sun on Mars at this moment in time is not strong enough to create any substance or forms of life on the surface. Venusians have been visiting our planet frequently through the vortex since the day we became capable of blowing ourselves to kingdom come. Their system is next to our system, a greater, higher system. A link with our two systems is through the vortex. If we blew up our world, would this influence their world, seeing as the two systems are so close? Earthly humankind could wipe out all life on the surface, including themselves. I do not think we could destroy the planet. The Earth has the capabilities to heal herself, without the aid of human life, to start this system all over again, from the beginning until it eventually works. That is if time gives it long enough to evolve. You cannot destroy the forwarding, advancing system, but merely slow it down.

Voodoo and Witchcraft

All true knowledge has been handed down to us from our Venusian gods. Some used that knowledge correctly, some, which was most of the people at that time, used it for power, greed, for selfish reasons only. To walk in the way of true knowledge would have been hard because they were of the minority. Our father God creates, and as being a part of God, we create also. We build for ourselves a home in the spirit world, brick by brick, formed from our thoughts, words or actions in this physical life. Some with Venusian knowledge have learned to communicate with spirit, to use that knowledge for material gain. Working with child-like minds and earthbound spirits. Not protecting yourself by not leading a good life, as the Bible says, will draw negative energies around you, and you will have had your lives and wasted them. The good have beautiful homes, productive, enjoyable work. The wasters of this world only put their energies to material things, and all that is material in this physical world will decay, leaving nothing as you enter the spirit world. Dark will be their world. A broken down, cold hut. One day he or she will pray to a god, which sends a light upwards; a guide will come to their aid.

Some do not even believe they have passed over into spirit, and there's a lot of confusion. Voodoo can help here. A coin with two faces, one good, one bad, in every belief system. The good side of voodoo is the communication of spirit to heal help and instruct on the true pathway. What creates good is in the influence of God. Black voodoo power, on the other hand, is for the uncultivated, superstitious, weak-minded human race with very little true knowledge and a total belief in voodoo doctors. If you believe you can be cursed then you will feel the curse of another. If you believe in a god that protects his or her little children, then nothing can harm you. Strong faith, strong mind. If God be with you, who dares to be against you? If a man cursed another man, and the man cursed walked in the way and protection of the Lord, the curse would do him no harm. The curse would go back to the sender. As for pins in dolls to bring about pain, a man must be primitive and simple-minded to use this method. The doll is shown to him and the needle is put in the doll, in front of the victim. A drug is given to weaken his ability to fight. He dies just like the voodoo doctor said, of fear, of a heart attack.

Witchcraft is the same – it has two faces. What you allow yourself to believe, is so. The mind is powerful, positive or negative. In Salem, two young girls having fits started up a myth that the place was full of witches. All who saw this great wonder then, not today, as we understand more, thought they were possessed by devils. They weren't and nor were they witches – just two sisters who had fits. Witches can practice astral projection. It is her spirit you see going through the sky, not her on a broomstick.

I have a strong faith in a creator, and nothing can harm us. Lead a good life; be protected by the light. See things as things are. A story of history cannot ever be a true lie; a small amount in any story is bound to be true. Searching for truth is a slow, painstaking, lonely pathway, like looking for a needle in a haystack. When a truth has been found this takes us upwards, going forward, advancing spiritually to a greater sense of reality in an illusory world. The spiritual high is worth the effort to get there. Your pathway is like a person charmed, no fear, no worrying about tomorrow. Working with spirit helps us to live a freer life.

We need to see cause that is negative before we put it into action and feel its negative effect. We need to know people beyond the frontal image they create for themselves, see a person as they truly are, have a higher sense of what is going on around you. You may see a person, the negative pathway they are taking. You give them advice, they ignore it, it is sad to have to stand back and let them go down that negative pathway. If you do not stand back you will get drawn down to the level of the person you have come to help. Try staying as a third person in any situation. Look upon a scene; do not be affected by that scene. Too deep an involvement in any patient's life or sickness can influence your own life. For anything you do with spirit, thank them for coming, and thank them for what has happened. This protects and will eventually teach you how to cut off from negative situations, as a doctor or a nurse does. To practice spiritualism you must have a strong mind, knowledge of what you are entering into and the time to develop your psychic gift, which we all have, although very few know they have them, and even fewer develop their gifts.

We have five senses. The average medium, which is usually a woman, develops one or two of the five, mostly clairvoyance or clairaudience. Seeing and hearing, we can develop any of the five. Five physical senses, the law of life says all that is physical has a spiritual counterpart, the spiritual sense formed out of the physical sense is what mediums develop. There is a sixth sense, the mind. The path I have desired to take. A path of knowledge and understanding is out there. To find it you need time and a strong dedication to your task. Jesus said in the Bible, 'Walk in this life, but do not be a part of it.' The Catholics, nuns, monks and priests spend hours in meditation. In a busy world we need to make time in our lives to relax and meditate. Mohamed, born to a rich family, asked himself why there was so much suffering. He gave up his rights, dressed as a beggar, went out into the world to find his answer. The more he travelled the more suffering in people he saw. Confusion became great in his mind. He sat down under a tree and spoke to his God in prayer saying, 'I will not move from this tree until I am allowed to understand all knowledge and understanding.' The people of the nearby village thought he was mad and felt sorry for him so they fed him. He taught of truth beyond anything that had been written in the old, wise books, the knowledge that the Venusians gods gave to the human race. He spent the rest of his life teaching that truth. From the day he sat down under the tree, to the true enlightenment within his own mind took ten years. Jesus entered the desert to receive that enlightenment.

I have studied sixteen years while employed as a security officer to receive that same knowledge. Most people have not got the time; all you need to learn is here in this book. You just need to read this one book instead of reading thousands to get to the same place.

Fables and Legends

There is no such thing as a miracle. Law rules all our lives, spirit law; if we live with the law, happiness, health and wealth would be ours. To go against law brings suffering and pain. We call a thing a miracle because we have very little knowledge of the law that brought this aspect about. Control is the key to a wonderful life. To control the base instincts keeps a person out of a lot of trouble. With kids you need control – a smack when doing wrong teaches them not to do that again. We all learn through pain; it is the only way to learn control. If you do not control your kids but allow them to run wild, with no knowledge of good or bad, then when they grow older society has to control them by drugging them with sedatives or sending them to prison. To be too soft with children you create a lot of suffering for that child. Teach them young; build a strong bond of love through parental control. A well-adjusted, hard-working help to society, not a hindrance, is a person brought up in such a way. Nature teaches through pain – if you put your finger into fire it will burn. Every time you do that, you will feel the effect until you stop doing it. It is better to be controlled within life by two loving parents than a total stranger doing his or her job, such as the police. To smack a child should never be done in anger, so try everything else first, and if those methods do not work then smacking is necessary. God understands the need to control children, so when giving form to the human beings he gave them a bottom, the only place you should smack a child. Hit across the head or any other part of the body will cause permanent damage. A bottom when smacked stings for a while, but you cannot permanently damage anything. Children should be controlled by the same sex as themselves, except in single-parent families.

Ley lines are channels of power from north to south and from east to west, holding the Earth in place; it would just break up otherwise. When two ley lines cross over each other, much power resides in this area. The English, in times gone by, built Stonehenge on two crossed ley lines built by the Venusian gods who settled down in England. The name of the ancestors of these gods was called druids. England follows a mother religious belief – a worship of nature. The last known practitioner of druid belief was Merlin the magician, in King Arthur's day. Druid knowledge formed into the white and black witchcraft religious beliefs. Witchcraft means woodcraft, and woodcraft means a worship of nature. The last of the druids was Merlin. The swords of King Arthur's day were made out of any metal they could find and they lasted a while, although they were easily broken, usually in battle. Merlin made a so-called magic sword, a sword that could not break, a sword made out of steel. Knowledge yes; magic, no. The lady of the lake was a spirit guide, seen mostly in the mornings, relaxed from sleep. Mist on the lake creates good shadowy atmosphere for spirit to work through.

England had many rulers, it was King Arthur's job to bring them together to work as one people to fight against their enemies from overseas. Merlin had great knowledge and he argued violently with King Arthur. King Arthur sent Merlin to the executioner to have his head severed. Merlin said, 'If I die this day so will the sun die with me. No light, you all will perish with me.' He lifted up his right hand and the moon came in front of the sun, cutting off the sun's light. 'Bring back our sun Merlin,' said King Arthur, 'and I will give you your life.' Merlin lifted his left hand in the air, and the moon moved away from between the sun and the Earth. The sun shone again. To this day, scientists call this an eclipse – not magic, just a lot of druid-cum-Venusian knowledge.

Dracula existed as Marcus the Impaler, a cruel, violent man who threw his prisoners from the castle walls onto spikes. He was not a vampire. Bats and only a very few species – snakes, leeches – drink blood to survive. Humans do not, except in festivals of the darker side. Life after death, in physical bodies, roaming the earth for souls, came from opening the caskets of supposedly dead people. There was no way in that day to be able to tell if someone was truly dead. They had no machines to tell them, like we have today. A lot of people were in comas, and woke up inside a casket, then suffocated because of the lack of air. When the caskets were open, it was often the case that the burial clothes were ripped and there would be a look of great fear on the now dead person. Matched with a superstitious people, believing that the dead came back to haunt the living, because of the wrongs the living inflicted on them, we can see how the myth of Dracula came about. Jesus brought two people back to life from a coma state. If a person is truly dead his cord that holds spirit to physical body would break. The only way for that spirit to be able to take physical life is to be born as a baby. Once the cord is broken, not even God could connect spirit with body again; even God for instance works within spirit law. Big Foot is an intelligent being. As other systems overlap our system, as in the Bermuda Triangle or the China Sea, creatures from their world enter into ours. If they come from a cold world their body would protect them in their evolutionary process by covering them with fur. Totally non-violent, they seem to be studying us as we try to study them.

What to do on Sight of Spirit

You may not be seeing spirit; you may be seeing images of past suffering in the memory of the land as it unfolds in front of you. If it flows by, with the souls seemingly lost in its action, they may not seem to acknowledge you. If so, that is not spirit, just a memory image from the past. If a spirit tries to communicate with you, ask a question with your mind: 'What do you want? Why are you here?' Then allow the mind to go quiet so the spirit can give their answer. It is possibly a lost soul looking for a little help and guidance. You should be in control without fear; they cannot harm you or come near to you without you allowing them to. There are many spirit levels, some good, some bad. If you get a strong, powerful Earth-bound one, fear would feed his or her energy, thus allowing them to be stronger in whatever they are doing. If you are not in control, the Earth-bound spirit will overpower your senses. When in doubt ask for your spirit guide, everybody has one. Ask God for guidance. Ask and you will receive. You should, before entering into a place that may be haunted, ask for protection from God or your guide close by. They know what you are capable of doing or what you should not be drawn into.

If spirit plays with you, negativity picks up more negative vibrations. They know your fears, they can play rough by using your fears against yourself. A strong mind, with a spirit guide at your side, can say, 'Hell, give me your worst, you cannot hurt me. My faith and my guide protect me. All is weak against the mighty power of the Lord.' If spirit of an Earth-bound nature gets too heavy, you are not learning anything from this experience; rather you are just hurting within these conditions. Light feeds light, darkness feeds darkness. Show spirit you are scared of them, and this feeds the Earth-bound spirit, making them stronger. Fear is weakness, strength is a powerful faith, in a loving, protecting father God. This draws a pure energy of light, cleansing, healing the land and the place, uplifting the spirit that is there. Some earth-bounds are so consciously deep into the attractions of the physical world, and drawn into the darkness, that they fear the light. The flow of energy from the light sends them to the place they have earned.

If an Earth-bound spirit will not go, say aloud in a strong, determined voice, 'Be gone, so the light and power of love can flow in this place,' and they will have no option but to go. To stop an Earth-bound spirit returning to the house, bless the house in prayer, sprinkling water that has been boiled once. Then offer this up to be blessed and put on an altar for at least a full day and night, and this will purify the water. Mix the water with salt that has been blessed and also been left on an altar. The water symbolises the spiritual power flowing from above, the salt symbolises the earth. Start at the top of the house, putting your finger into the blessed salty water. Mark the sign of the cross over windows, doors and all outside opening vents. Pray in every room, sprinkling the water, 'Oh Lord God Almighty, Maker of heaven and Earth, send down your healing power of light to this room, taking away and not letting return all that is negative.' Light lives and breathes in light. Darkness cannot live in the light, and darkness can and will one day evolve towards the light. Do every room, leaving the corridor in front of the front door until last. The house will be emptied of negative vibrations. All that was in the house has now been drawn into the corridor. Open the door, sprinkle the water about, say the same prayer as in the other rooms, then say in a strong voice, 'Go spirit, do not return.' Put the sign of the cross over the front door with the blessed water.

If still concerned about this spirit, put a small wooden cross over the front and back doors. Always try to communicate at first. See if they will tell you what they want or why they are there. Nothing can harm you unless you allow them to harm you, through inviting them into your life or through having fear of them. Your home is your home. Spirit belong in the spirit world. Do not draw Earth-bounds towards yourself by opening up to spirit or playing about with spirit as with the tarots or any instrument of communication that is not done with sincerity. Have knowledge of what you are doing, with a prayer for protection and a guide close by, if possible, before you start.

This is a form of training in mind control for directing the mind upwards so as to connect with only the highest. It is a course in meditation and how to search within your own mind for a truer and better you. If you simply play about with spirit under the influence of alcohol or drugs, without knowledge of what you could be dealing with, it is no game; fire burns, and you will get mentally and spiritually burned, allowing your mind to enter into spirit blindly. Earth-bounds are closer to us than what higher spirit is. Open the doorway to spirit without the protection of a guide, and you are opening your life to any low Earth-bound spirit who wishes to enter. Those with psychic gifts and who use them for gain of great amounts of materialistic wealth lower their level of communication with spirit in the end; their only communication will be with lower Earth-bound spirit. They know very little about anything positive. If you have a power, use it properly – to help other people and to try and prove to all we come into contact with, that life goes on forevermore. Abuse the power and you will lose it to the lower spheres. Possession by spirit is very rare, only happening to people who are superstitious, weak-minded or who are mentally sick. Overshadowing is a more common aspect. You would have to draw yourself down towards the spirit wishing to overshadow on the lower levels, by drugs or alcohol, which opens you up to the level you are at, or what state you are in during the attempt to communicate with spirit. Being overshadowed by a guide, like in a spiritualist church, brings through stronger spirit messages, powerful healing and deep sincere philosophy.

To know yourself, then to know your guide and then trust in the guide to overshadow your body, to leave someone else in control – that is trust. Mediums of today do not go that deeply. They enter a development circle, once with a mild type of gift, more nos than yeses. They enter onto a platform giving the general public a bad display of their gifts. Development within the same group should carry on, and there should be no end to development. Some gain a moderate level of spiritual communication, then give up with the developing circle, as if they know it all; then they wonder why their gift is not improving. This is also why we cannot fill the spiritualist churches. In a church one power is matched by spirit in communication. In a group many powers must work together to raise their senses. If a medium experiences things they do not understand, in not understanding they worry about it. Being in a group, different people on different levels, some will have passed your way so as to be able to correct what is wrong in your development. Church-developing circles are pretty safe as compared to developing by yourself.

In a lower Earth-bound state there was a spirit gentleman who liked his drink while in physical life. It was drink that eventually took him into spirit. Looking into the physical world he saw a drunk, just like himself. Like attracts like. The spirit overshadowed the drunken man in the physical. The spirit could now enjoy the feeling of alcohol getting his physical instrument to drink more for the spirit's enjoyment and thus sending the physical person into spirit with him. It is the same as a murderer overshadowing a murderer – you have to be that way to draw the spirit to you. Be good to yourself and to others as this brings a strong protection in physical life.

Some people, either through drugs, alcohol or electric current treatment, find themselves opened up to spirit, and this causes confusion. They go to a doctor who is no wiser then the person is. The doctor will put them on drugs which opens them up a little more. As the situation gets worse then it is into hospital for ECT, which totally opens you up. Before drugs, a person needs counselling or sending to a counsellor. The counsellor should have some spiritual knowledge, so people who have opened up to spirit can be advised on how to control themselves and that it is only the next step in human evolution. Knowledge removes many a fear. Our doctors have not got the time, nor money from governments to pay for this type of treatment. Their way is to drown their senses with drugs, and they will go away with a supply of their latest fix. They are not living or learning from life, but stagnating. They then accept and live with the fact that they are sick. Nobody expects anything from them. They get to enjoy their challenging lives, becoming very lazy and eventually becoming totally dependant on that drug.

A female friend of mine, when young, fell in love for the first time, very deeply. The man was wrong for her, and they eventually broke up. She was heartbroken, and suffered bouts of depression. Her doctor put her on anti-depressants, and twenty years later she cannot even remember the name of the man who had caused her so much suffering when young. But she is still suffering from addiction to the drug her doctor gave her.

Fear

Give a person knowledge and they will have no fear. Fear is of the unknown, such as death. But to have knowledge that life goes on is to know you will meet again. It is a happy time, when a young spirit who has travelled a long journey arrives home – his true home in the spirit world. This knowledge for a true believer should take away any fear of death. Going through the pain leading up to death is pain you have created for yourself. If you eat the wrong foods, do no exercise, smoke and treat all the people you come into contact with badly, then when your passing is a painful one, who can you blame except yourself, when physical life is coming to an end you can prolong that pain and time of suffering by not letting go of the physical. Then you must accept in faith this next journey within the light, and that the last journey is at an end.

All that is physical will decay; only that which is spirit lives on for eternity. The powerhouse of the physical body is the mind. What you think is what you are. If you think good positive things, you create good positive vibrations. All power of thought will attract to itself the same as itself. If you think negatively you draw negativity towards you. A man afraid of being attacked creates for himself negative vibrations because of these fears. Negative vibrations go so far into space, before they come back at us. So if you fear being attacked, you will one day be attacked, brought on by your own negative thinking. The same happens with a positive thought. Fear of animals, snakes, mice, rats or anything is either the fear of the unknown or created by a happening in your past. Not just in this lifetime, for we have lived in some form of physical life, millions of lifetimes. If you are scared of anything, that thing must have hurt you at some point, or how would you know to be scared of it?

Most fears form in childhood, and some are imaginary, only within the child's mind. Released into the mind of an adult, held onto, not faced, and so not conquered, the fears multiply. Face fear, get to understand it then get rid of it. We carry a lot of unnecessary baggage around with us. Some fears are real, like things that have happened that are too deep to work out logically on your own. You need a good listener to speak to, bring all your fears of the past out into the open – a trusted friend or a doctor. Then release, and never give mind to those subjects again, except to look on them as a point of learning, something not to be repeated. When you look back, you look as if looking at the actions, and the sorrow created with these actions is detached, not evolved, as if looking on the life of another person. Otherwise looking back would be painful. We all live, we all die; in death we pass onto a greater, more conscious state. As we pass from one life to another, the passing may be violent, creating deep scars on the subconscious mind, that come forward into the mind of your future life. Fear of falling, for instance, may mean you fell to your death in some past incarnation. Claustrophobia may mean you died in a cell or you have been in a coma, buried alive, awoken in the coffin. Everything in all that exists has a cause. Find the cause and the healing power will flow to the pain of that effect. If used properly, hypnosis will bring about a lot of healing, not just to the physical mind but the spiritual mind also, the true you.

Everything that has happened since originating from a one-celled organism which became a human being is registered in our subconscious mind. In taking the mind back to past lives we can understand the cause of the effect we are going through in this life. Until a fear is faced it will always keep on coming back. Face it, understand it and then get rid of it. Fear of spirit is mostly brought about by wrong thinking. We follow like sheep to a shepherd instead of thinking for ourselves. We follow other people's ideas because they seem better – all that is, is clothed in negative vibrations. We had knowledge; our Venusian brothers and sisters gave us this knowledge. We abused that knowledge, and now it is hidden away from us. The truth is still there for all of us to see if we desire to see. Rip away the negative and all that is left is the positive. To say God is good is to say that there is no part of him or her – God being a collection of the two in one being – that is bad. The power of God creates; there is no destructive force within him or her. God does not kill. Now you will be able to read the Bible with this statement in mind, with a deeper and more truer insight.

When we leave physical life, accept the memory of loved ones we have left behind, who we will one day meet again. The rest of physical material life should be left behind us. Attachment to things, houses, wealth, intoxicating sensuous pleasure of the body – all can draw us down to the Earth, as badness we have created will keep us earthbound. We need to go back to our spirit homes built brick by brick with the good we have done. Bricks decay with bad actions, thoughts or words. Negativity is strong within physical life, and physical life decays. All that is negative will decay. Life is light, positive action a never ending advancement of the soul. When we grieve, which we need to do to release the sorrow and pain of losing a physical loved one, we allow the healing power to flow. Some become so engulfed in the sorrow of the passing of a loved one, drawing toward themselves a build-up of negative vibrations, not realising these vibrations are actually cutting them off from the one they desire to be close to. Spirits come very close to us, and try so hard to make contact. When entering into spirit, your senses of your spirit body are heightened; they sense and feel things more deeply than a person in physical life.

The first conscious thought would be of all the negative feelings that you have for a family now grieving because of the love they feel for you. This opens your mind, and all your senses to what is true and what is false. Nobody judges us, we judge ourselves. If the pain is too great for that spirit being, all he or she needs to do is call on the light, and any belief system in a greater power and understanding than themselves. If you have no belief system you have no knowledge of the light. When you suffer, suffering is personal, within. Help comes from within, and if you do not know that light then who will give you that help, love and understanding? Ask and you will receive, do not ask, and you shall stay in the dark, karmic world you have created for yourselves. There is no need for this when there are a thousand hands reaching out with love in their hearts, giving freely their service and creating as much good as they can in our lives. Most people live in a darkened, closed-in world where spirit cannot go. We need to reach out towards spirit asking for help. Help will come to anybody who asks, whether they be in the physical or in spirit. Spirit cannot enter or influence anybody's world unless you invite them into your life. We are the masters of our own lives, and spirit can try to guide us upwards. Relax with music, blank your mind from material things, think of a loved one. If practised regularly you will feel the touch of your loved one.

Grief is emotion running wild; be calm, relax, allow that healing power to flow. A physical person entering into spirit will be pulled by two forces, the desire to rise towards the light, and the desire to stay on the earth plane because of the sorrow his or her passing has caused. If you wish to help spirit, let them go to a place of peace for a while to work out the lessons he or she should have learned on Earth. Believe that every one of you, good or bad, Father God and Mother Earth, love all their little children. Spirit hands are outstretched to help any person who asks. Upwards you will go and the time, pain and suffering is up to you. If you learn the lessons well the first time, you will live easier and experience an upward spiritual advancement. The Earth was created for the pleasure of the human race. It is the human race that has made Earth what it is today.

Numerology

In numerology you have 0–9. In the alphabet are 26 letters; put 1 to 26 next to each letter:

A	B	C	D	E	F	G	H	I	J	K	L	M
1	2	3	4	5	6	7	8	9	10	11	12	13

N	O	P	0	R	S	T	U	V	w	X	Y	Z
14	15	16	17	18	19	20	21	22	23	24	25	26

This method will help you to find your lucky number, possibly a lottery number, together with a lot of faith in the power of God. Everything affects everything else. Some numbers you are drawn to, others you push away. Take your own personal numbers, your name. We will use mine as an example:

S	T	E	P	H	E	N
19	20	5	16	8	5	14

S	K	I	N	N	E	R
19	11	9	14	14	5	18

$$19+20+5=16+8+5+14 = 87$$

$$19+11+9+14+14+5+18 = 90$$

$$87+90 = 177$$

If higher than 9, add together $1+7+7 = 15$

Add together again: $1+5 = 6$ Lucky numbers:- 6, 87, 90, 15

A lottery has only so many numbers. If a number is over that which is required, add them together to create a lesser number. You can take the numbers from family names, addresses, and dates of birth. All have personal lucky numbers, the numbers to register the spirituality of a person. Then there is one to seven. Seven days in a week; in six days God created the Earth, and on the seventh he rested. The Age of Aquarius is the seventh day; we live in the last days of Pisces, the sixth day. Numbers one to seven are for those who have succeeded: for instance, six represents a receiver of knowledge. One denotes that a person is being more materialistic than spiritual. I am a six, a receiver of knowledge in a time of understanding.

Numerology is a science. A lot of the great pyramids have a library of numerology, leading to greater understanding. When the human race has advanced and is ready for new knowledge they will then be able to read and understand that new knowledge. If a lady wishes to find out if a man is suitable, she can find her own personal number, then find out his number. One to seven are the numbers for love. If the man's number is higher, only one number up from the female, he will be the more dominant. If the female's number is one above, she will be more dominant; both should be good working relationships. If the number of the male or female is more than one point up or down from their partner then the greater distance between the two numbers suggests a lesser chance of them making it together.

In the Bible, numbers have a lot of meaning. Seven is God's day. We are the third planet to the sun. There are twelve tribes of Israel and twelve disciples symbolising them. Forty, either in days or years, was the time of great learning. The Jews spent forty years in the wilderness learning the law, trusting in God and suffering for their past negative actions, breaking down bad karma so they were good enough, pure enough to enter into the Promised Land. The only one who did not make it was Moses. Forty days and forty nights it rained for Noah, giving him time to build up trust in his maker. Jesus spent forty days and forty nights in the wilderness taking in the true purpose of himself. To me, forty in the Bible represents a time of inward struggle to reach a beautiful, spiritual awakening. The time of any of the seven days of creation is about two thousand years. We are nearing the end of Pisces; Pisces started at the birth of Christ. Pisces is a fish, which lives in water, which pertains to a time of baptism. So if you can calculate one day, it's possible the rest will be about the same. The seventh day, God's day, on that calculation, will be cut short because by that time the planet will be too hot for human life, so we will all go to Mars, and become the gods of the Martians. Paradise on Earth and in the spirit world will last a thousand years. Then the planet will end in fire as the Bible predicts. Let us hope that when we go to Mars we will have learned enough lessons to do a good job, as the Venusians did before coming here.

Television

Every family in the western world has one. As a nation we spend more time watching TV than anything else. The behaviour patterns of most people have been learned from television. Parents are too busy to teach or control their own children. Engrossed in materialistic gain, they have no time to talk to them. 'Be quiet, watch the telly, stay out of my way.' Too busy making another pound that they don't really need. People put more importance on a new car than their child, the child they brought into this world. A child is for life, a thing some parents forget. It is the parents' job to control and educate children to be helpful, hardworking members of our community; it is not society's job. Thousands of pounds of our tax are spent locking up youths in an attempt to control them. Law is law; nobody should be above the law, whether it be child, priest, policeman or politician.

God in his great wisdom has created a book that covers all situations. This book is the Bible, which most kids or adults have never read. Television is their Bible, teaching them how to live in society; robbing, beating people up to gain power or profit is the done thing in this television-taught society. The only crime most people understand is getting caught. From very young, we are desensitised towards violence, drugs and sexual abuse in everything from cartoons to war films. The darker, negative side of all our personalities should be controlled; television brings it out. When we see police raping, thieving and getting away with it on TV, most people are conditioned to believe that the police are like that in reality. So a greater gap is created between the average person and authority. A twelve-year-old knows how to inject drugs and has knowledge of where to get them by their last year at school or possibly even sooner, the way our system is going. Children should be taught the true facts about drugs and sex. If we do not teach them, then they will pick up false ideas from TV or their friends on both subjects. Their friends probably know as little as they do, and these two subjects should not be learned about via experimentation, as many lives are ruined this way. All drugs need to be legalised because people are rarely given help until they fall foul of the law, by which time it is usually too late. If a person has to get the drugs from a doctor then society knows how many people need help. Cannabis should be sold like alcohol is, as it cannot harm you. It is better for you than alcohol as it relaxes you, and therefore there is less chance of you becoming violent. Tobacco is the only poison when smoking cannabis. If cannabis could be made in tablet form this would help many people cope with the stresses of everyday life. Depressed people are given pills which are far more dangerous over a long period, often provoking bad side effects – and they are given these by their doctors. Cannabis is also a plant; nothing is added to it. All other street drugs are mixed with anything the pushers can get their hands on.

You can only control something when you know who the users are. Less crime equals a more controlled society. Drug counsellors should be in the schools. Drug pushers, with the exception of cannabis pushers, should be given heavy prison sentences, with prisons being prisons and not holiday camps with all the luxuries of life. More re-education needs to be carried out in prisons. Parents have a responsibility to their children. Televisions, newspapers and radio all have a responsibility to the society that they broadcast to. If society works together as one, everything in life is possible. Clear the evil, vile substances off our streets. To legalise would cut down on needle-infected disease. Prostitution should also be legalised, put into houses and off the streets. There should be regular health checks, and they should carry health cards. Any prostitute working on the streets without a health card should be sent to prison for at least two years. They should also pay tax and stamp on their earnings, like any other working person. Attacks with weapons or even simply carrying a gun, should carry a very heavy prison sentence. Prison should be a place of rehabilitation, changing, if possible, the way a person thinks, so that they may become an active, responsible person within our society. How a person has changed by the end of their sentence should be on a psychiatrist's report before they are released. If it does not seem like they have learned their lesson they should not be released. This would be only for crimes of violence or sexual offences. Prisons should be run like a business so they pay for themselves. Why should I, an honest, hard-working, taxpayer, have to pay for prisons? If a person is in a high position in life, a judge, a police officer or a care worker in our hospitals, their sentences should be doubled because they have abused their positions. This also goes for priests, with immediate removal from their position for life. Electronic tags should be attached to the more severe criminals, the time for which they wear them being dependent on the severity of the crime. Sexual offenders against children should be tagged for life. Jesus said in the Bible, 'Suffer the little children to come unto me.' Anybody who brings suffering of a sexual nature upon another will not be forgotten, in this life or the next. Besides murder of another person or yourself, sex abuse is the second worst crime, whether it be against a child or an adult.

When films are shown on TV concerning child abuse, battered women or drugs there should always be a helpline number screened at the end for those that are suffering these situations within the family. Only sex shops should sell pornography; all other shops should be banned from selling such material. There should be heavy fines for offenders if a shop sells cigarettes, solvents, alcohol or fireworks to children. A thousand pounds fine on the first offence, prison on the second. If we do not get tough on crime, life will not be worth living.

People live in fear; this is no way to live. People need to change the way they think. Nothing is free; everything has a price, whether the price is financial or physical. Every man should look after and provide for himself and for his family, not live off the society they live in. Women should work, if they are not looking after kids or older family members. To work is to be active; if you are not active you will have time on your hands, causing boredom and low self-esteem. There is work out there, look and you will find. Life, work, opportunities of life will not come knocking on your door. A man should be made to work even for dole money. Most young people just sit back and let life pass them by. They take drugs to relieve the boredom, just the same as alcohol, and in the end the children suffer because they have to go without.

It is time that the world faced up to its own responsibilities, personal and global. You hear them say, 'Why has this or that happened to me?' But most of the sorrows of life we bring onto ourselves. Nothing restricts us from rising to any high in society we desire, except ourselves. I am a person of average intelligence, a security officer for the past sixteen years. I looked at life and asked myself why. Now I know, and as proof I have written this book. We all have talents; most do not get used. Jesus told a story of a master who was to travel to another land. Before he left he called for his three servants. To the first he gave five talents (money or skills), to the second three talents, to the third one talent. On the master's return, the first had doubled; the second likewise. The one with one had hidden it in fear that he may lose it. The servant with the one talent had that talent taken away from him. The others profited, and he lost everything. Do you, reader, use all the skills that God has given you? Use them or lose them. On returning back to spirit, as we all will do one day, will we have profited from our life or have we allowed life to pass us by? It is up to us, as individuals, whether we rise or fall, find happiness or sorrow. Look at your life today: is there something missing? Find out what it is, and then go for it, in an honest God-fearing way.

I once lived without hope, my life full of fear, blind to my own possibilities in life. Now I am free, nothing worries me. We create our own chains that bind us. The past is gone and should only be remembered for the lessons that have been learned, so as not to make them again. The future is built up of what we do today. Nobody knows what the future holds – not even spiritualist mediums – for life is cause and effect. A medium can only tune into the effect of an event we have already caused. So worry, if you wish to worry, about today only, for it is the only time you can do anything about. Being in a physical body is like being in a valley, with mountains all the way around us. We can only see the mountains; spirit life is like being above the mountains, and thus the sight and understanding of spirit is greater. We can through meditation tune into the spirit of ourselves, giving us a greater insight about what life is all about. It is a waste of time, shouting, moaning and groaning about life. Answers come in the silence, in prayer or meditation. Take time in the silence from your busy world, and a happier more confident spiritual person you will be. Stress is the killer; all disease is activated, brought into being by stress. Love yourself, and in so doing, others will love you. I get up every morning, look in the mirror and say, 'I am the best, and am able to cope with any troubles of the day. There are no problems in life, just challenges which I will conquer.' So it happens; how we think is what we are. To relax in water is good. It reminds us of the womb, comfortable and peaceful. The Romans knew this; they put a great importance on bathing.

Sleep

We need to sleep to recharge our spiritual batteries, the spirit being that which is truly us. So sleep recharges our physical bodies, which need food and water to survive. Our spirit needs to be fed with and from the life force that flows down to us, from our father God. The way we negatively treat our physical bodies restricts the flow of the life force. It is very hard for power to enter into deep gross matter, as our world is built up of. So at night, as we sleep, the spirit leaves the body so it can get its full supply of energy, which it has earned. The spirit cannot fly away; it is connected to the body by a golden cord: a stretchable golden cord which cannot break until the time of your passing into the spirit world. It is a little like the cord that you were connected to your mother with, when you were in the womb. If you had consciousness of this recharging of the spirit within sleep, you would have consciousness of spirit, and be then able to learn how to move and communicate within spirit. Some are able to do these things, in astral projection. The more relaxed the body, the deeper the sleep, and the deeper the sleep, the more conscious of spirit your mind is. Relaxing music, sounds of nature, birds and animal noises, water lapping onto a beach, bird sounds in the background, and any relaxing music that you can hear, without thinking about, are all ideal in achieving that deep relaxing sleep.

Thinking is action – you need a relaxed body, not an active mind – and via sleep you will eventually be able to perceive spirit. The healing power flows at these times, recharging and refreshing the physical body. Sleep has five levels of depth. The first would be light sleep, when one is partly asleep, partly alert. A most perceptive time for being taught from tapes, relaxing tapes or educational. The second is dream sleep. Your eyes are cameras, clicking pictures all day, then played back as a film of the experiences of that day. Sleep is a place to release all those positive and negative actions, to sort them out and try to put them in some order in physical life, as a safety valve for emotions, if you like. Then we enter into deep sleep, when the spirit rises above the body to recharge, and it is in deep sleep that loved ones on the spirit side come close to us. Finding it hard to communicate with you, they send little messages of guidance in the dreams formed on this deep level. Questions are answered; we ask the question in light sleep, and receive the answer in deep sleep, in dream form. Running away, being scared of someone chasing you in a deep sleep dream, means you will not face up to a certain situation in your life. You are running away from it; stand and face your fear of this situation, then sort it out. Falling seems to be a common occurrence in dreams. When the fully-charged spirit enters back into the physical body this can give the dreamer the sense they are falling.

Flying in a dream is when your spirit has found out how to travel in spirit. Think of the sensation of travelling with the mind; imagine where you want to be, and you are there. It beats buses and cars, and does not pollute the atmosphere either, and you are firmly connected at all times by the cord to the physical body. Flying dreams are usually followed by a strong intense dream, remembered when awoken. Sometimes if lucky you will dream of entering into a garden, meeting all the loved ones who have passed on into spirit. Everybody should keep a pen and notebook at the side of the bed, to write down dreams upon waking, before you forget them as the clouds of earthly problems cover your mind.

If you do not believe that all I write about happens, it will not ever happen to you. Believe, then, in spirit and your loved ones you will see again, while still in a physical body. Spirit are reaching out to us on Earth. Because people do not believe in spirit, this forms a negative barrier across their frontal mind. Much true knowledge, help and comfort, is unable to penetrate the physical mind, because of wrong religious thinking. Sleep is the mind, in spiritual action. The less stress we enter sleep in, the deeper we go. A lot of sleep is taken up in working out the worries of the day. That relaxing time that we all should have, tunes us to spirit. When we are sick, it can often have been activated by stress, and if we have no belief in a healing power, we are on our own. Spirit or the power of spirit cannot help us. To believe is to always be able to tune into that wonderful, healing power. 'Believe and be healed,' Jesus said in the Bible. 'Your faith has made you whole.' Whatever spirit gives you, be it in dream form, or in any other form of spirit communication – test them, and if what they are saying is true, that is a spirit to trust. Jesus said in the Bible, 'A good tree bares good fruit. A bad tree bares bad fruit, know them by their fruits.' As we enter into a new age, a golden age, the Age of Aquarius, workers are short, and so is any channel of thought sent towards the light. You can work with spirit: many spirit helpers are there to help you achieve your desire. As we enter into the last days of this system, moving towards a natural disaster of the kind that the Earth has not seen since Noah's day, there has never been a more important time for knowledge to flow from spirit through physical instruments. Knowledge and the practice of this will take us forward to that next stage in the spiritual advancement of human beings.

The World of Today

In our world today, everything we eat, drink or breathe is poisonous to our body system. Water is so polluted that it is wise to boil it before use. They catch a lot of bacteria in testing. A lot of poisons are getting into our water systems: pesticides sprayed onto crops, filth we put into the atmosphere which permeates the water as it rises from water to steam, coming back to earth as polluted rainwater. We breathe a fog-like filth, which is in the air. Pesticides are a fairly new chemical product and nobody knows what long-term effects it will have on human, animal or plant life. There has not been enough time to test long-term effects. The only good food is that which you grow yourself. The water board makes the water we drink clearer. They have very little money to do anything with, because of bad management, so we all have to suffer. Many pipes have had very little maintenance for the last hundred years. Now very old metal pipes are becoming contaminated with flakes of rust entering into our water system. Let the water from a tap flow for a while before drinking it. Lorries, cars and factories are also polluting the air we breathe. Look at the pollution we inflict daily on ourselves, through smoking and drugs, which are also in our food, placed there by the manufacturers to add flavour or colour. If we do not smoke we take into our lungs about as much as five cigarettes a day, anyhow, just being in the presence of other people who smoke. There is no place in hell that is as dark as the colour of a lung of a smoker. The great amounts of polluted water we drink are also present in beer. It is a battle inside our bodily system to be able to cope with so many new poisons. Many diseases are new, not formed out of history.

The age we live in is failing us. We need to go back to natural farming, without chemicals, meat of living animals, any one of them our younger brothers and sisters. We have lived many incarnations within physical life. In some of those, we would have been animals. Even animals today are pumped with drugs, injected into them to make them bigger and more profitable. They are fed on the carcasses of their dead family. We think we are human, yet we are still barbaric savages, on account of how we bring so much suffering to the animal world, by eating and enjoying. If only we gave them some life on a farm, free to follow their own natural instincts. Instead, they are kept from birth up to the day they are brutally slaughtered, in cages so cramped, so small that they can hardly turn around. Some are tethered in small, dark, cramped rooms, the stink of death in every breath they take. Meat damages us physically; and damages us spiritually too. People kill animals to eat, but a state far worse than that is when an animal dies for its fur or ivory to please the vain people of this world. Animals and birds are also killed just for the pure pleasure of killing.

There is a spiritual plane of consciousness, created just for the souls of suffering animals and birds. Trained animal doctors are there to help and heal the mental pain and suffering of those souls which have been inflicted by the human race. What we do to the animals comes back to us as disease. Many human beings are passing over to spirit with diseases that have formed out of the animal world, because of our abuse of that world. We destroy all that is beautiful. One day a child will say, 'What is a tiger?' and the parent will show them a film or a picture of a tiger taken many years ago when they still roamed. There may well be a time with no animals, due to the rate we are killing them off, and wiping out many species that will never return.

If the human race cannot put a value on a thing, that thing becomes a pest and is wiped out. 'What is the point of anything existing if I cannot use it or make a profit from it?' says the human race. People look towards the future and ask what it will be like. To create a better or a worse tomorrow, we must live with and work with our younger brothers and sisters. We must share our space with them, enjoy their beauty. Like all that is, if we misuse the Venusian knowledge or the knowledge of nature, as humans have a tendency to do, we disobey the spiritual or physical laws that rule our system; not man's law, God's law. Man's law bends and is sometimes even forgotten when money changes hands. Whatever it may be, if we abuse it, we lose it. Shoppers of today need to read and try to understand what they are buying. It would help if manufacturers could label the ingredients in much simpler terms, so that a person may choose. In today's world, we go on trust that any food is exactly that: good for us. The average person does not know what half of the ingredients mean, composed as they are with ten-letter words, with no mention of whether they do you good or not. Only the most intelligent can look after themselves, as they know what they are eating because they can understand the language. The rest just have to hope they are not poisoning their own children. If anyone seriously studied what goes into our food, they would make a huge impact on the industry. But only if the book was written in a most simple form.

Great Britain

What is so great about a country that conquers and takes what it wants with no thought for the inhabitants of those conquered countries except to turn them into slaves or wipe them out, as England has done wherever it has been, be it Africa, India, Australia or New Zealand. The colonists were chasing after gold and other precious metals, using the inhabitants to dig it up for them. Many places that had not known great diseases suddenly found millions of their land's inhabitants wiped out because of bugs brought in by their conquerors. So great was the cloak of darkness for natives that it was hard to find the truth, since their version of the truth was stripped from them and replaced by new force-fed religions. Religion is the control of many, through fear, and a worship of idols. True knowledge was lost to them, and they worshipped images of a lesser age, until the rock or stone that the image of a God was moulded or chipped out of became the new idol. The stone image became the God controlled by a few wise priests, for voices, for collecting money for this God and inflicting punishment to whomever dared to think they were wiser than this God, even though made out of stone. Hardship through their religious oppression directed them straight into the hands of white British men selling their own personal version of religion. Taking what we wished, we slaughtered all those who were against us. We forced on them our beliefs, so they came out of a strong controlled religious belief and went straight into another one. A white man, to many races, was thought of as God.

They ruled their inhabitants with pain and sorrow. There was no 'great' in Britain; the only thing that made British conquerors was that they had guns as compared to bows and arrows as in the Native Americans. The British tried to perpetrate against the Native Americans what Hitler tried to do with the Jews: genocide. The Native Americans were a peaceful race who lived off the buffalo. They used everything on the buffalo, for clothes and weapons. They were a very religious people, a nature worshipping people. They believed in a god who came in a ship, just after the great flood.

They knew how to tune into the spirit world, and had vast knowledge in healing potions. They were a peaceful people, whose minor conflicts were soon forgotten after they took a smoke of the peace pipe. Archaeologists dug up a Native American graveyard, and found a pipe embedded in clay. The pipe under microscopic examination showed a black, tar-like substance. This substance happened to be cannabis, and that is probably why they had very few battles. This was before the white man, when the Native American was at peace with his god, his land and within himself.

So easily bought was a Native American, too trusting of a white man baring gifts – bracelets and mirrors for the women, hard intoxicating firewater, it just sent them crazy. They enjoyed the gifts; they wished for more and told the white man, their so-called friends, about gold and where it could be found, just trying to please their friends. Gold was not as valuable as the buffalo to a Native American. You could not eat it; it was not a good medicine and provided no clothing. The Native American mined gold for the white man; the white man did not know where to dig. There was a seemingly endless supply of gold. The white man became greedy, and their refrain soon became, 'No firewater until you tell me where to dig for the gold.' The Native American responded in kind. The white man tried to take the gold by force, with use of weapons. This created a war between white man and Native American. The white man took their gold, took their land, and tried to starve them to death on government-run reservations. The money was coming in from the government to feed and clothe the Native American. Officials of these reservations only spent a small amount on the Native American, keeping most of the money for themselves. A Native American, dead or alive, was worth a few dollars. It's a long haul to carry one dead Native American to a stockade where this type of dealing was done. It was easier to just cut off his scalp. Yes, the white man invented scalping. The Native American just copied what the white men did; they were not the savages. Every promise a Native American made, he kept, but the white man broke every treaty he ever made with the Native American. A white man dares to call a Native American a savage. The white man is the savage. One glorious victory against Custer was achieved by the Native American nation, becoming one nation. If it had occurred earlier it could have changed history.

Reincarnation

There are twelve signs of the zodiac. Each sign is a pathway and that which you travel as a male, you will also travel as a female, giving twenty-four incarnations from first consciousness of your self to true spiritual awareness. Every sign has its lessons and we must, if we wish to rise and develop spiritually, learn all the lessons necessary, but not always in the order they are in, which is earth, fire, water and air. Depending on what needs to be learned, a person enters physical life under a certain sign. An Aquarius, which is a high sign, means that down one pathway they have already travelled. There are many pathways, and for some lessons the person who is an Aquarius could take a lower level path to learn the lessons of that level, indeed, if it was necessary to learn those lessons.

There is good and bad in every sign, no level being greater or lower than the other; all have lessons to learn. If a man mistreats a woman in this life, in his next lifetime the man will be a woman, mistreated by a man. Law is perfect; you cannot get by God's law. You create the cause and you will feel the effect either in this world or the next. We get away with nothing. What you do to others will be done to you. We are born to learn about our weaknesses and we look for strength in others to take for our mates. If two people, who live together, could not learn anything from each other then there is no point in the relationship between the two. What attracts two people is the fact that one can learn from another. A person can only achieve a few things in life on his or her spiritual pathway: become a stronger and more powerful light, and go back to spirit a flame of glorious light, greater than the flame you left spirit with. A person may have many opportunities and take advantage of none of them. This person would stagnate, making no progress one way or the other. Stagnation comes from lessons not being learned: you will be back in physical life until those lessons have been learned. Get it right the first time – it will save you from a lot of pain. Once a lesson is learned then it will not repeat itself in physical life.

Around our bodies is a cloak of light which is called an aura. All you have ever thought, said or done is registered in the aura. There probably are many auras, many levels of light. The physical aura deals with how you use or abuse your bodies. Illness shows up on the physical bodily aura before it shows on the physical body. If doctors were spiritually tuned in enough to be able to see another person's physical bodily aura then they could cure the sickness before it manifested itself on the physical. Anything is possible if it's a positive action backed up by spiritual energy. There will be a time when spiritualist healers will work with, and as part of, the medical profession with faster tried and tested healing methods, that have been passed on from the beginning of time from Venusian intelligence. All true knowledge is Venusian. It is the spiritual aura that registers all about us. Every negative aspect creates a slightly darker aura. Every positive good vibration makes the aura shine many different colours during the passing from the physical to the spirit. The light of the aura is the level of your home earned on how you have led your life. The opposite of this is to sink deeper down into your own karmic debt.

Karma

What a person does comes back to them. That which is good helps your spiritual advancement. Whatever negativity you create for yourself you must work out that negative where that action took place. If created in the physical that negative should be worked out in the physical. In the Bible it says 'Thou shalt not kill.' To take any life is wrong whether it be an animal or a criminal or insane person. To sentence someone to death is the hallmark of an uncivilised society that cannot handle what that society has created. Sending the problem to the spirit world, and letting them sort it out is not the answer. Rather, it is re-education, finding out why they committed those crimes in the first place. Help them to understand themselves, for they need help. Sending them to spirit does them no good at all.

From the sun to Pluto there are many levels of consciousness. The further away from the sun the darker the conscious level. Every level of thought draws towards itself the same as itself. Light draws light, darkness draws darkness. It is very hard and would take a greater amount of time to work out negative karma. You cannot be truly tested on whether lessons have been learned or not, except in physical life. Physical life has the greatest good and the greatest evil on one level. It is the best and hardest school of learning that there is. Suffering comes from not learning. How long we suffer is up to us. The purpose of existence is to evolve from the darkness to the light. On entering spirit with so much negative karma, the spirit world on that level that they had entered onto would darken because of their own actions. The executioner who pulled the switch or carried out the lethal injection, they would also be classed as murderers, carrying the blood guilt of the person they had executed.

One of Adam's sons in the Old Testament killed his brother. God put a sign upon him so anybody finding him was not to kill him. If God does this to a murderer and his or her justice is true, as we are taught from childhood to believe, should we be taking the life of murderers against the way God handled people on that negative level? Do we have more knowledge of justice then our creator? What can we do with murderers, as the Bible teaches us? Do not kill them and cut them off from society, but try and teach them a better way. 'Justice is mine,' says the Lord. Anyone who takes their own life is a murderer. With God's law all aspects of that incident will be taken into account, including why you did as you did. Physical life is important – you can learn faster and rise quicker and spiritually higher in the physical world as compared to any other level of thought that there is. To take your own life is to cut your learning off before it has finished. Whatever problems a person has so as to commit suicide they will find the problems they have to work off in spirit are far greater because of that action, when our physical life is about to come to an end. If we are going through great suffering of the mind and pain of the body, this is the result of how we have lived in the physical.

God is merciful; just give yourself into his or her care and attention. If this does not work and you wish to end your physical existence then you can ask to die. If you can give your permission in writing or in front of a witness who is not a member of the family, you will not suffer for this action. In the spiritual, you will have suffered enough. Jesus came to us in love, healing the sick and raising the minds of the suffering, to show us human beings the way home, into the light. All the good Jesus did was immense, whereas some materialistic people now are capitalising on God, making vast profits through fear and ignorance and holding power over the masses, as some religions do. Their time is short. The truth will stand, and the rest will fall; they will have their reward, just not the one they were expecting.

In the Holy Land on the spot where Jesus was supposedly born now stands a chapel. Before you can enter an expensive entrance fee is asked for. In the streets people sell to the gullible, religious folk who have saved for years to make this pilgrimage. They sell pieces of cloth supposedly from the 'actual' robes that Jesus wore or selling off the five nails that pierced his body, a piece of wood from that same cross. It is amazing that people hand out great amounts of money for these things. They have already earned their place. Religion is not a business. What did Jesus say to the merchants in the temple? 'God's house is a house of prayer you have made it into a den of thieves.' Some churches of today come under that category. All churches have knowledge and most of that knowledge is covered over with ceremonies and old worn-out doctrines which have through many years lost their true meaning. All had their purpose on many different levels.

You need to pass through some belief systems to be able to evolve to a greater and truer system. All belief systems are steps on a ladder. We live in the age of Pisces, which started at the birth of Jesus. The year 2000 brought into being the beginning of a new age, the Age of Aquarius; that which you know at this time, is one truth the Piscean understanding of the age you are in. New, greater knowledge and understanding comes with the next stage of the Earth's spiritual evolutionary advancement. Out of the old follows the new, one formed out of the other. This system, and all systems, are continually evolving. Our purpose for our existence is to evolve from the darkness to the light. There are many levels of understanding, like school, with infants, junior and senior. You would not take an infant-school child and put them in a senior school; they would not cope. Religions are on many levels, all with their own special truths, evolving one out of the other. A person could stay in one belief system, but, as they are not evolving, they will eventually end up stagnating spiritually. We are an evolving species, spiritually and physically. Knowledge of the spirit world provides gifts of healing, as well as relaxation of mind and body. These things are all around us. Do not have too closed a mind, have an open mind. Everything evolves and if the minds of those of the mainstream religious beliefs do not evolve then they will stagnate.

Satan's time has the number 666. Six as a number is the sixth day, the seventh being God's day, the day of enlightenment for all, a thousand years of peace. This time of peace, which will come, was told to us as the symbolic story of Adam and Eve. Paradise on this Earth has never been. Paradise is a dream of a future time. We are entering into that time now. Although light is being formed out of darkness, before the light can penetrate our system, peoples' minds will have to change and until they change a positive existence will never spring from our negative one. Suffering will build up. The sky gets the darkest just before the dawn. But have faith, God looks after his own. The first six of the three represents the breaking down of false beliefs, only allowing that which is true to stand firm. So negative, wrong ideas will be no more. It will be a cleaning, sorting out time to reflect on where we have come from, and what new and greater possibilities will open for them in the future. The second six represents the financial world, in which stocks and shares will collapse. There will be changes in temperature, the loss of crops. Like a snowball rolling down a hill, things will get worse, get bigger, and a breakdown will occur in that system. The third six represents a breakdown in our society in general: a breakdown in morality. It will be a time to go back to basic Bible teachings – the Lord knows how a person should lead their lives. It is a way that has been tried and tested since time began.

Astral Projection

We are spirit. We take on the physical body, which is clothing for the spirit, in order to learn lessons in the physical world, the school of hard knocks. Everything will grow old and decay. When our bodies are worn out, we discard them. Then we enter into spirit, our your true selves. The physical body is of no use any more; burn it, bury it, it does not matter. It is like an old coat that is worn out. Our bodies need food and water to stay alive. The spirit needs to be fed too from the life-giving force that flows from God. To receive this power, our spirit, in deep sleep, hovers over our physical body, recharging itself for the next day. If you could not enter deep sleep for some reason, and were made to stay awake nights and days for a long period, your spirit would get weak, and this would eventually show up on the physical. The more spiritual power that flows down to you the stronger you become, because power strengthens the immune system.

To have consciousness of being out of the body and being able to travel in spirit is called astral projection. To do this, lie on the bed listening to some relaxing music. Say three times to yourself with determination 'I will, when I have entered into sleep state, have full consciousness, with clear memory of all I see or encounter in this sleep state. When I awake I will recall all details of those matters which will be beneficial to my physical and mental being.' At the side of the bed you should have a paper and pen to write down all inspiration on the point of awakening. If the desire to project astrally is strong and practised regularly then there is no reason why you should not be able to do it. If you are weak-minded, on heavy drugs, scared at what you are doing, or merely is playing with spirit, then you will get nowhere. The stronger the desire, the more chance of your spirit rising with full consciousness out of your physical body.

You will always enter back into your physical, for you are connected to your physical by a golden cord, just like the umbilical cord. The further you allow your cord to stretch away from the physical, the stronger your consciousness forms in the spirit. The closer the spirit is to the physical the stronger the senses are within the physical. Spirit can only see the physical world when close to the physical body. For any spirit to see or communicate with the physical world they need an instrument or medium to work through. Whatever you do where spirit is connected, ask beforehand for protection or a guide to show you the way. The truth is out there. The human race was given knowledge. Firstly the female then the male. The male is still abusing true knowledge. The Bible is written on two levels: one level for anybody to understand, the basic way of leading a good, positive life. The other level is symbolic, for the deeper, sincere reader. God is good, and in him there is no darkness at all, so any violence towards anybody that is supposedly sanctioned by God is totally wrong. Look for more earthly, natural disasters. Colour, like Joseph and his multi-coloured coat, represents the spirituality of that person. Nakedness represents knowing a person, nothing hidden. Everything follows a pattern, that pattern is true, logical. If something is fanciful, hard to understand, then look for a deeper meaning, asking of the light with a quiet mind.

Death means the end of a situation, not *the* end. As one situation finishes another more glorious adventure is about to start. When one is born into the physical one dies in spirit. When one is born into spirit they die in the physical. The only true meaning of the word death is a place of very little light. We all need some light to keep us in existence. Until the time when we, as a spirit being, cannot learn anymore from physical life we will continue entering into physical life. When we have learned all we can learn we then live in spirit, not to be given physical life again. When the seed in man enters the eggs of a woman, giving growth to an embryo, for that growth to be taking place spirit must have entered the embryo. No spirit, no life, no growth. If the embryo is damaged and there is no doubt about how great that damage is, which could produce a mentally deformed child or endanger the mother's health, then the mother, after receiving all the medical information, could agree to an abortion, as early as possible. Anything else is murder. But what life, you must ask yourself, would that child have had if it had lived. Quality of life is the important issue. If an embryo is healthy and is aborted, then that is murder and you shall suffer for that action. A lady I know aborted an embryo when she was sixteen and had much parental pressure to abort. Later, all her friends were getting married and having babies; she could only look on with desire in her heart for a child. Something went wrong with the abortion, and doctors later told her she was barren because of the damage. She spent her young life looking after other people's kids, never having one of her own. One day she entered a church; the light flowed through the stained-glass windows. 'Oh God,' she said. 'You gave me a miracle of life, and I aborted that life. I am truly sorry and I understand my wrong actions. I am so sad Lord at not having a child of my own.' One month later she was pregnant. Ten years after she aborted her first; ten years of suffering. What we reap is truly what we sow. There was a man involved.

Blood

As the Venusian gods mated with earthly women – a higher conscious level mating with a lower conscious level – it created a molecular disorder within the blood, which we call AIDS, although it has been known by many different names. A man should not mate out of his cycle. If you make love to an animal, that is out of man's cycle. The Venusians needed to mate with earthlings so as to develop earthlings mentally and spiritually. AIDS nearly brought the children of Venusians to an end. Roman emperors, pharaohs in Egypt – all were very sick, many crazy. The Venusians saw what danger mating with earthlings was causing so they left the earthlings and moved up into the mountains. Roman rulers and Egyptian rulers died out in a very small space of time. Children died without even reaching their teens. Blood, if properly tested, should be fine. It is an individual choice and risk. This law about not taking blood was made before humans could test blood. The more we evolve in tests, the purer the blood. Most of it is very well checked.

Forgiveness

The law of life is cause and effect. When Jesus spoke of forgiveness of sins, he had just healed someone, connecting sin with sickness. If a man was a cripple who had worked out bad karma and learned the lessons that crippled him in the first place, he would then be ready, spiritually and mentally, to be healed. All healing to one degree or another is beneficial to people. True healing heals the spirit, which heals the body. Unless a person is ready for that true healing, enlightening power, they would not be able to receive that high level of power. We get out of life what we deserve. Jesus channelled power to the sick, and if the right conditions are not there true body and soul healing cannot take place. So for Jesus to be able to channel the correct amount of power to heal a cripple, the cripple must be ready, and have earned the right to be healed. When the healing happens a man's sins have been worked out or forgiven, if you wish. Nobody can, by saying prayers or giving donations for the church, absolve themselves of sins. God cannot even forgive sins; God has to go by the universal law just like we have to do. Law is ultimate, in true justice. All that is, is influenced by law, and no one can get away with anything, not even God. Why should he want to? After all, God is evolving just like us.

There are two types of power that we can draw around us, positive or negative. Positive allows you to rise spiritually, negative keeps you closely drawn towards the physical world. If a person does you wrong do not go out for revenge, but hand them over to the proper authorities, or you will only draw yourself down to their level. If you love, you draw love to you, if you hate you draw hate to you. You need, so as to rise spiritually, a working out, a breaking down in your mind of all that is negative. If you do not do this you will attract that which you hate towards yourself in the next lifetime. If a person hated blacks, Spaniards, any country or people, then in the next lifetime they will be what they hated. Life is like a coin with two sides: if you hurt others you will discover how it feels to be hurt by others; it will come back to you. So it is best to forgive people no matter what they do. Carrying their negativity around with you is destructive to *you*, not the one who has harmed you. This is hard but true forgiveness is the forgiveness you give to each other. To forgive sends a light from you to the person you have forgiven. This light raises the consciousness of that person, creating in them a stronger awareness of themselves. This is good for the progression of that person, but not for that person becoming aware of the negative actions. This causes mental pain and suffering of their own making until they move on and change their lives for the better. If you do not forgive, you end up carrying a lot of emotional baggage around with you that belongs to other people. All the negative thoughts given off build up around that person like a thick cloak of negative vibrations. To forgive is to bring hope and light towards that person. To hate takes you down, and also pushes the one you hate deeper and deeper into darkness, where the light shines only weakly. All that a person does positively feeds back to themselves and their spirit being. The spirit becomes brighter, feeding a spiritual group, a family unit. All that is positive in you recharges and reenergises the group as a whole. The brightness of this group strengthens the plane of consciousness that it is on. That plane of consciousness, now stronger, lights up the plane above it. This action goes straight up towards the God. That is how God is reenergised – from the light that shines from every soul.

Where is God?

They say God is in all things. We see God's beauty in the birth of babies, animals and nature in spring. We can perceive God's power, but where does this power come from? All that is, is light formed out of darkness. So total darkness is the lowest level, total light at its highest level is the sun. God is the spiritual power behind the existence of the sun. God made man in his own image. Worship God in spirit and in truth. God is a spirit, as we are also. We are physical and so is God, the sun. In the conscious level of God everything would be perfect. No man, no woman – just a being which is the highest level a human can achieve. Masculine and feminine merge into one being – not a body as we know it but a conscious living energy built of the purest light and colour. There are millions of these minds working together to try and raise the level of spirituality – all levels try to raise the level that is just below. Our sun, our God, is being drawn towards a greater sun, a greater God. The sun at this moment is weak, and it needs to draw Mercury towards itself so the sun can engulf Mercury, creating a couple of million years of sun energy, when the sun will shine stronger and will create a warmer Earth, also creating a rise in spiritual consciousness all over the Earth. A person would know him or herself as they are, but they will also be conscious of a higher state of mind, two conscious levels that can be perceived at the same time, the higher and lower of oneself. It will be a time of madness, of confusion, of many lost and not knowing which pathway to take, and instead, blindly heading towards any of the man-made new age cults.

Spiritualism in the west is the only way. The heat of the sun will activate the core of the Earth with a greater influence. As the core heats up, this will create a build up of pressure at the surface, leading to volcanic eruptions and earthquakes. With greater heat the ice caps will melt, creating floods, and freak weather conditions. Spiritualism as a belief system has not progressed enough, but will progress enough when needed to be able to help the many lost souls that do not understand what is happening to them. There are many conscious levels and we are in just one conscious level, that which we call physical life. To rise and to enter into a higher conscious level we must be able to perceive the level we are on and the level we are heading towards. This is what is happening, stuck between two conscious levels. This will happen every time a soul wishes to rise from one level to another. Darkness is strong, in this day and age. The light has to counteract the darkness. You could have risen to that next level at any time in history. The knowledge was there for a person to know how to rise. Now you have no choice, the sun will shine and the spiritual power behind the sun will shine. The race as a whole will go forward, no matter what.

For God or spirit to communicate with the physical earth, they need an instrument or channel. Spiritualism needs teachers to teach true understanding. The more people that open their hearts and souls to God helps to strengthen the power of God that flows to this Earth, making the transaction between physical and spirit easier. Gifts of spirit are not really gifts, and much time and energy is taken to develop them. All the gifts of spirit have their place within the spiritualist movement, as has knowledge, the foundation of any belief system. What the congregation wants the congregation gets. It keeps the churches full. Communication with a loved one, a message from one who has passed on to the brighter side is a good tool for proving life after death. The people want messages all the time, giving no time for spiritual knowledge and understanding, when this Earth at this moment in time needs to be able to take that next evolutionary spiritual step forward. Three spirit teachers in tune with our world at this moment in time are: White Eagle, Silver Birch and Zodiac, to give knowledge so as to be able to take us forward as easily as possible. There are many great teachers and these are just the ones well known in the spirit movement to bring forth the golden age, the age of enlightenment, the Age of Aquarius, the seventh day, the Lord's day.

The spiritualist movement needs a positive, strong belief system which they all can agree on and teachers to teach that system so as to help all those lost souls who are opening up to spirit. We used to believe that, but now it is not enough – we need to know. People are asking questions that the priests of the mainstream religious beliefs cannot answer, and so the churches are emptying. Spiritualism will grow stronger once it has learned to teach. Is spiritualism the true religion? In the Western world, yes; in the Eastern world, Buddhism.

The Devil

As God is a collection of like minds which retain their own personal individuality, so the devil is a collective group of similar-minded people. The Venusian gods came down to this Earth like a parent wishing to keep an eye on their children, the earthlings, bringing with them their knowledge of how to live a happy peaceful life, to try and raise them spiritually upwards towards the light of God. To take forward this primitive race they took Earthly women for their wives. The inter-bred children with knowledge of a Venusian and the spirituality of a child had a strong desire to control and conquer all lesser earthlings. They even tried to destroy their own Venusian parents or the gods in physical form at that time. They believed themselves to *be* gods. Some allowed themselves to sink so low that the attraction of the physical world bonded them to the lower spiritual planes. Not all mixed-race children went the way of darkness. Much good was done with advancement in the sciences, the medical world, astrology and maths. A time of great knowledge was also a time of great suffering. It was not just those of mixed-race who broke spiritual law – so did many of the earthlings following in their pathway. Now these corrupt Venusian inter-bred, spiritual children live in the lower planes, the planes closest to the Earth. Some of the gods, who were pure in heart, felt sorry for their misguided children and the state earthlings had got themselves into by following their ways. These Venusians, with pure hearts, true children of the light, stayed in the lower planes to help and guide their younger brothers and sisters who had followed the wrong path. They have control of those levels. These true Venusians by their own choice live on the lower levels.

'Satan' is a group of teachers trying to help their younger wayward brothers and sisters to rise towards the light. The fall from heaven to hell was their children falling from the true way to power and greed. Upon his death, Jesus entered into the lower realms, and his light was so strong on that level, although it had been weakened after the crucifixion. The light of Jesus lit up the lower planes giving to the spiritual beings of that place a ray of hope, a ray of pure light, lighting up their world, and directing their minds towards the light. Jesus in himself was the seed and the light. His teachings give all physical human souls the chance to go home to, and be one with, the light. He entered into the lower planes and lit them up with the power, the light of himself. Not just giving hope for the physical world but also the spiritual world, hope for all, whatever level they may be on. If we had all the knowledge the human physical world possesses, and knowledge was a bottle of milk, all of what we know is but the cream on the top. We have still got a lot to learn.

Freedom of Choice

The higher you are the more you work with the universal, ultimate justice of the law of God. Spirit cannot enter into a situation until that person allows them to. This is a law of the higher side. The lower spirit beings will be drawn to any physical that is close to the spirit's own personality. A drunk could attract a drunk in the spirit world, but a physical human being would have to lower themselves down to the level of an Earthbound to be influenced by one. Everybody has to go by universal law. Law has been tested time and time again, for it is the best way that physical human beings can learn their lessons, so they evolve to become a higher being, the easiest possible way.

You were not meant to suffer: look around, all is for your enjoyment, the countryside, the ocean. Behind every negative there is a positive waiting to unfold into your life. Look for the light within the darkness in any situation. Create for yourself a positive attitude; believe in yourself as an individual. All of us carry the spiritual Christ light, the seed in each and every one of us. Look for that light in others, look at what they are good at, not always at their faults. Law is there to help us. If a parent says to a child, 'Do not touch fire, it burns,' and if that child puts its hand into the fire, it will burn. We should take notice of what our own spiritual parents say or we will get burned in other ways. Law was given to Moses, then we were given greater law through Jesus. To go with God's law is a positive, productive, happier way. If we break God's law, we suffer. After we have gone through a time of suffering brought onto ourselves we turn back to how God says it should be done as he knows what is best for his children. We will turn back when we have had enough suffering and we will then realise that God was right all along – if only you had taken notice, that would have stopped all that unnecessary suffering. We do not think of the effect of any cause until afterwards and then it is usually too late. God tries to guide and protect his spiritual children, just like your physical parents would wish to protect you. God or the power of God never turns away from anyone. Nobody is ever truly lost; eventually all will find their way into the light. Through our own negative actions we cut off light from ourselves, and God out of our lives. A person has a choice whether or not they decide to take alcohol or drugs. While doing so, if they commit a crime, it is their choice. They choose that state of mind so they should take responsibility for all actions. Time in the spirit world is not that important. If a soul is going forward and learning, that is all that matters. To get from one plane of thought to another could take one lifetime or a thousand lifetimes, it just depends on how quickly or how slowly you learn. Time is a physical aspect not a spiritual one. The only time spirit needs to work within our time, is when they come close to us and our world. A person's greatest enemy is themselves. Why cry out, 'Look at my suffering'? You will stay suffering until you find the cause that has created that suffering. Look for the light and the way out of that situation, and then change it for the better. Hard, yes, but it can be done with a little positive thinking, nothing is impossible. We are the masters of our own lives and if anything else becomes the master of you, get rid of it. Try and sail your ship, your life, the body and soul of yourself towards the light, your future home.

Rising Thoughts

Spirit cannot communicate with the physical. Spirit communicates with the spirit of the medium. The spirit of the medium then tries to impress that thought on the frontal physical mind so the medium can give that message to the people. Know yourself, and in knowing yourself you will be able to tune into your own spirit self. Your spirit self will tune into the minds of other spirits, connecting with the spirit mind of a person, linking with the physical mind of a person. We send out thoughts depending on how strong or weak, negative or positive they are. Is their influence and effect on all that is around us? Mostly it is with those who have a direct connection with us. Positive thoughts rise upwards towards the light. The more positive and strong the thought, the greater that thought merges within the light. Whatever spiritual consciousness, that thought reaches the spirit level. The spirit on that level would match the power or thought sent up, with power and action of thought sent back to try and bring about positive thoughts for that person in the physical world.

If you think of one who is suffering, that sends down the healing power to that suffering person. This works on every level: the more you give out, the more will come back even greater. Be careful what you wish for – you may just get it. Good thoughts are healing thoughts; bad, negative thoughts do harm to people. A person who gives off negative thoughts all the time to other people will eventually find all that negative energy directed back at him or her. All that is physical will decay; spirit is what goes on forever. Positivity raises the mind; the negative clouds the mind and darkens the aura. It also keeps you close to the Earth plane because of strong negative desires. So how you think creates for you a future spirit life, a spirit home. When your time comes, as it will do for all of us, to pass on into spirit, the golden cord which keeps physical and spirit together breaks. The spirit rises from the physical, the brightness or darkness of the light of your aura is the plane of consciousness that you have earned a place for yourself, and it is your future home.

Cycles of Life

Life is built up on cycles, and our systems have twelve cycles, twelve signs of the zodiac, every sign a cycle, the cycle of human life – baby, child, teenager, adult and so forth. There is also the cycle of the seasons. God is good and does not like to see his children too tired. So lessons need to be learned. After a lesson has been truly learned this creates an extra burst of spiritual energy flowing towards that person. Because if a lesson has been learned, the person who has learnt would be more positive in life, have a better outlook. They would send up positive thoughts, have a happier life, and that is why the extra energy flows. After a time of trouble, peace comes for a while. Then there are more lessons to be learned, and more peace at the end of it. Some stay in negative cycles, but with a little effort they could be in that next phase in your life. Negative thinking creates fear: of today, and of moving on. No matter how dark life can seem there are many spirit persons' hands wishing to help you. Help any of his suffering children. Ask in prayer and you shall receive the answers. All human beings have the capability to draw, and to tune into the energising productive powers, which are all around us. Everything has life, from a one-celled organism to the Godhead. Life is first brought into being as a thought from a higher level. That thought form would form in nature as the spiritual power behind a flower. The life force, that gives life, so as to enable that flower to grow.

After the task has been completed by this energy, which is called elementals, the energy flows back into the atmosphere and is one with that energy. Some elementals evolve and take on matter such as rock, coral and then fish, evolving along the same path as the human race, developing one day into human form. Everything is linked to everything else. Some pathways of physical evolution do not evolve far enough to take on human form, such as birds. Their souls take on the forms of the little people, fairies, gnomes and the rest. As the elementals had a task to do – give life and growth to the plant world – so do the little people have their job to do working within nature. Air, water, fire and earth all have elementals creating their natures by evolving through the four elements: air and water a more feminine side, fire and earth a more masculine side. Some fade away into the mass, some evolve further, taking on the form of little people, like mermaids at sea. If you think of the pathway of physical human beings as symbolised by a tree trunk, then the little people are a branch of the main trunk. Once any form of energy starts to evolve then it will continue to do so until it has reached the highest state it can reach. If an energy reaches as far as it can go in physical life and cannot return into matter then that energy's progress and development is spiritual. They learn from human beings by working close to the Earth plane. So fairies are spirit people formed from the bird world. Children, who have natural psychic ability, can see into these natural planes and see all the little people busy at work, working within nature. Most children are sensitive to the fairylands and spirit planes; parents usually reject any conversation on these matters as being silly, a stupid realm of the fantasy, rebuking their children whenever they mention these things. A child learns to keep certain things to themselves as their awareness takes on whatever is going on around them. As they grow, with all the problems this entails, they cut themselves off from all the other worlds that are around them. Maybe they will open up to spirit later on in life. Once we are able to see or communicate with spirit, this is a gift we have attained and which can never be lost. It will wait in the background of your subconscious mind for a time, before becoming active again in physical life.

Everything has an aura. Fairies are little innocent young souls; the light of their aura we see as wings. Only the birds have wings – angels and fairies have beautiful shining auras. To travel is to think where you want to be. Do it and you will find yourself there; there is no need for wings. Fairies are so much freer to express how they feel, by movement, which to us seems like flying. Fairies eventually evolve into young angels, giving their lives in service for the physical plane. Everybody in physical life has a guardian angel. In the quietness, think of your guardian angel. They can give a lot of help, guidance and understanding to their children in the physical world. If only we would quieten our minds and listen to those gentle angel voices.

It is not always the fairylands that children tune into. There is a plane of thought formed for the purpose of educating children that have passed on before their time. Lessons on this plane are learned not forcefully but in play. Sometimes physical children can tune into the plane of spirit children at play. With Aquarius comes the laws of that age, in which all humans have the right to be treated as equals no matter what colour, religion or status they have in life. Thin or big, sick or healthy, able-bodied or not, good-looking or ugly – all are equal in God's eyes, all travelling down that same pathway. Animals are our younger brothers and sisters. If we do not look after the animal world we will lose it. Higher levels of consciousness in physical life feed off our younger brothers and sisters, creating a plane of consciousness that consists of great suffering because of how humans treat the animal world. From the slaughterhouses come these suffering younger souls.

In the spirit world there are doctors and nurses to try and bring our minds forward, to try and heal the suffering and pain of physical life. In the animal world, in spirit, there are many human helpers to help heal and bring forward the young suffering souls of the animals. This plane is unnecessary, and when the human race learns to live with animals as younger brothers or sisters then this plane of thought will fade away. Meat, although it seems good to eat, creates blockages within the flow of the life force, which we all need to exist. Power cannot flow as easily as it should, bringing about many bodily ailments. An apple has a life force within it; you do not just feed the body, you also give food to the spirit, the life force being in whatever you are eating. Dead meat is exactly that, and has very little life force, and blocks the spirit instead of feeding it. When developing psychic gifts, meat eating creates many problems with spiritual communication. The mind needs to break away from a thick layer of negativity before it can reach the positive, the spirit.

Freedom is a word shouted out from the rooftops, but the

only freedom we have is the freedom when we have been released from our own lower selves and have been given the opportunity to rise. Lessons of life are all around us. Look at a flower, see how it grows, formed out of a seed in the darkness of the Earth, like the Christ light in all of us covered in the darkness of negativity. The sun shines on the seed in the ground. The spiritual sun which is God shines on the seed within every human being. Because of the light, spiritual or physical, the seed in the ground or in humans grows, evolves, pushing itself out of the darkness, towards the light. Slowly but surely, at each step, changes takes place. Bulb to stem, stem to bud, until the flower breaks through in beauty and elegant fragrances. So does the spiritual seed, in the human being, spiritually open up towards the light. Nothing that is permanent comes into being quickly. Slowly but surely opening up, that which is physical, decays. Once accumulated, anything that is spiritual, is eternal, and cannot be lost.

Homosexuals

In this country, England, we need stronger privacy laws. What goes on between two consenting adults is nobody's business except their own. We spend far too much time bothering about who is doing what with whom. We need to be more aware of our own lives, not the lives of other people.

We formed from one power, then we became two powers, male and female, and we will eventually return to the one power. We are spirit, in the clothing of the physical, and whatever sex we are depends on what lessons need to be learned in that lifetime. As spirit, we have the capabilities to be either male or female, and which of the two is the strongest is what a person's physical form will take. God is the highest level of consciousness of a male and the highest level of consciousness of a female, intertwined throughout many centuries of evolution: the ultimate form. The masculine and feminine, the one being, the highest form in physical life, is a dual-sexual reproductive being. The Earth is over-populated, and now women have the choice between family life or a career, fewer babies will be born. If the strongest influence in spirit is partly masculine and partly feminine they will enter physical bodies as partly male and partly female. They are born the way they are and have the right to live their lives as they wish. All will one day become aware of the masculine and the feminine within the one body.

In Mexico, a UFO spacecraft crashed. Because of the breaking down of the cooling system in the craft, most of the aliens were burned, but the parts of the body that survived were beautiful – a beauty beyond our concept. Male and female, the difference was only slight – some were slimmer, some had bigger foreheads. The brain is a muscle: the more it is developed, the more space it takes up. Everything has a purpose, including homosexuals, who open up the physical awareness of a dual-sexual being.

An Evolving Universe

God is the spiritual power behind the sun. God is still evolving. The law that God goes by is the same law as we evolve through, the same law, but a deeper law. One day, through the experiences of many lifetimes, we will be a part of that God force. We are a part of God now, in a very young and immature way. No matter how high we reach spiritually, we cannot lose our own individual personality, not even within true God-consciousness. One day this planet, Earth, will get too hot for us to live on, so we will go to Mars, which by then will be a developed planet, yet at the Neanderthal level. We will be their gods as the Venusians have been ours. The God light is in all of us, growing brighter and becoming stronger, with every physical thing we do.

All good or bad deeds, in thought, word or action are registered in the aura as light or as shady parts. However great or low your aura shines at the passing from this physical world to the next, that is the light of the conscious level of spirituality you have earned for yourself, your future home. You enter into physical life to experience, and hopefully to learn from those experiences. If you enter back into spirit with a greater, brighter flame than when you first left spirit to take physical life, then you will enter spirit to the sound of rejoicing, with many spirit souls there to greet you – guides, helpers, your guardian angel, plus all your family that have passed on before you. If the light of your aura is the same or lower than when you first started this incarnation then you will have failed to take the opportunities that the law of karma set out in your life. To waste your physical life to work out bad karma in spirit is a sure process, but a slow one. It is quicker to learn your lessons in physical life. No matter how low or shady the plane of thought you find yourself on there are always helping hands from higher levels trying to reach down to you, once you have directed your minds towards the light and asked for help. Without the light, a soul is truly lost in a world of their own making. The word 'dead' means a place without the light. It is up to us as the individual to make the choice, to either rise or fall.

Advancement

Everything evolves – the Earth and all that is on it. We learn lessons, becoming stronger, slowly dispersing all that is negative within ourselves. We will go through wealth and poverty, health and sickness. You do not need to suffer as greatly as you do; with relaxation, meditation, and the light, you can empty your frontal mind so spirit can register thoughts on it. This will help you get through physical life better. Nothing can influence your life, whether it be in the physical or spirit, unless you allow it to. On any plane of thought in spirit, like draws towards like. On the physical world there is a high level of good, but also a very low level of bad. It is a good testing ground, a sound school of learning. Spirit can travel on lower planes on their own, but cannot enter any higher plane than their own without a guide. If you entered on a plane of thought which you had not earned the right to be on, all the knowledge, the sensitivity of that plane, would be too much, like overloading a computer with data beyond its memory capabilities. Guides take souls up to higher levels to show them what they can reach at some future time. It also keeps an image in the mind, a goal to strive for; you see the higher plane and you wish to live there. Lessons need to be learned so you can earn that right, like a carrot in front of a donkey. We have all sinned and have to work off sin. Accept your own karma, and always remember, it will all work out for the best, for God's in charge.

The End or a New Beginning

Those who worship God and live their lives accordingly will get through any future disaster. God looks after his own. Spirit will direct anybody who wishes to be saved from any destruction, to a safe place, if we listen in silence to the guidance of the spirit people. Like the ostrich, we as a nation have a tendency to hide our heads in the sand hoping future disasters will never occur. To survive, all nations have to work together as one toward a common cause – the Mother Earth. We have to advance to the level of understanding of all people. As brothers and sisters within the light or as a scattered world, it would be hard to see if we, as a scattered nation, would survive the next earthly spiritual step. We need each other to survive in this time of suffering. Great acts, amazing achievements: the spiritual light will shine on all the Earth. A flood of love will fill all the people's hearts, helping others to survive this destructive time. There will be one belief system, a mixture of Buddhism and spiritualism, as east will meet west on religious thinking. Do your best, that is all you can do, then God will do the rest. If all earthly human actions fail, we still have Venusian gods, waiting to be accepted amongst you so they then can show us how to cope with this destructive age. UFOs are our friends, maybe our only lifeline. We as earthlings have a long way to go; let us hope we have the time.

The Dream

I dreamed a dream of a land of peace, where violence was no more, where a soldier would throw down his gun in order to pick up a banner for the Lord: a banner with a few simple words, a few simple words of love. Then I knew that I was with a heavenly host above. Then I heard sweet children's voices, singing praises to their heavenly King. Then an older, wiser spirit-person came my way, a person whom I did not know, although he seemed to know me. He said, 'When you fell, it was my hands that lifted you. I have always been close by you and always will be, because I am your spirit guide. This land you see today, you cannot stay in. You will return to this land one day. Until that time, tell everybody who will listen what you have seen, in the land of the heavenly dream.' A cloud shrouded my sight, and this land faded into memory. Falling down towards the Earth, back to my bed, I heard my mother's familiar voice, 'Get up, get out of bed.' I got up, looked out of the window to the sunny, blue morning sky. A bird was flying higher and higher into it. I wonder if that bird has been where I have been, in the land of the heavenly dream.